Content

The Author

Guido Bruch studied economics in his hometowns of Duisburg and St. Gallen (Switzerland). Directly after his studies he moved to Munich to work as a management consultant. Especially on the recommendation of savings banks, regional banks and tax consultants, he helped medium-sized companies to finance special events such as investments, company acquisitions or reorganisations by means of business plans. As a business angel, he was involved in the startup VisCheck (mobile image processing with a focus on quality assurance) in 2015. The already existing interest in digitization and industry 4.0 became stronger and so he became aware of the new robots at an early stage. With MRK-Blog.de he runs an important website in german language, which reports neutrally and usually with videos about the Cobots, another common name for MRK (human-robot collaboration).

Together with the automation specialist Omron, VisCheck has developed the first system for reading production screens. With the information digitized in this way without complex interfaces, a robot can then intervene in the production process to correct the situation if necessary. The system is presented in the chapter "Solution to the shortage of skilled workers". The Japanese company Omron makes its robots, its AI and its Edge Computing available for this purpose.

His experience with the needs of medium-sized companies and his analytical thinking trained in drawing up business plans help him to advise SMEs interested in robots. Bruch offers an inventory incl. recommendation at a flat rate. 50% of the consulting costs for SMEs can be subsidised by the state. In addition, Bruch is available for lectures on MRK and prepares market studies, in particular also for the German area.

Guido Bruch is married and has two minor children.

A small definition

MRK (human-robot collaboration) or Cobot is understood to mean human-robot collaboration. This means that the robots are not only as compact as lightweight robots, but also so sensitive that they cannot endanger other people during normal operation and as long as they do not hold sharp objects. This also applies if a person turns his back on them. In addition, newer MRKs are very easy to program, some of them offer pre-programmed work instructions in the form of apps and therefore have a never before known flexibility.

Important

The book contains numerous links. So that you do not have to enter them, you can request the slightly modified manuscript free of charge as a PDF file. This means that you can read the book and simply click on the screen. The PDF is intended only for the book buyer, a passing on represents an offence against the copyright. buch@mrk-blog.de

Offer to buy

The author will be happy to take stock of the possible uses of robots in your company. This is done at a flat rate (2,000 € net in D/A) and a few hours on site. You will then receive concrete recommendations in the form of a report. 50% of the consulting costs for SMEs are subsidised. He also prepares market studies and lectures on robotics. guido.bruch@mrk-blog.de

Despite the greatest care, errors can be excluded. The author offers apologies and assumes no liability. Before a possible purchase, the robot data must be checked again independently - also because there may be modifications.

Version 6.0 October 2019

Foreword

This book has been written in particular for those in charge of smaller companies and even larger craft enterprises (SMEs) who suffer from a lack of staff, who may not find trainees for whom full automation is not useful or too expensive and who have realised that it cannot go on like this.

This group of people should be convinced that a successive robot application - be it in the unmanned third shift or with somewhat larger series - pays off and that its application is not costly. Instead of complete solutions, smaller isolated solutions are proposed - often under the leadership of the youngest employees. As the boss of the largest robot manufacturer said: "Often the children of the entrepreneurs are the ones who want the Cobot. With the new robots, trainees can also be recruited, for example, or later tied to the company. Make your youngest, if they are halfway smart and technically adept, your robotics experts!

In order to show the potential of the new, inexpensive (from € 15,000 - incl. accessories/workstation certification rather approx. € 30,000) and easy (partly via app) to operate robots, the reference to a press release from the beginning of July 2018: The german Gigaset manufactures (cheap) mobile phones in Germany for the first time again and this at Chinese costs. With an investment of 400.000 € now 2.000 mobile phones/ week can be produced 70% automated. If the robots were to be fully earned directly in the first year, the production costs per mobile phone would only amount to €4. This example shows why robots usually pay for themselves within less than a year (∅ 195 days according to Universal Robots). This applies in particular if the aspect of employee fluctuation is taken into account (costs of searching, re-learning) as well as the reduced error rate. Even with the purchase of a "luxury variant" the hourly rate of a robot will still be less than 3 euros. In addition, a robot can be used to process orders that would otherwise have been rejected due to a lack of capacity if there were a shortage of personnel. In this way, contribution margins can be generated that would not be achievable without robots.

In the following, after a short historical review, decision aids for the application itself as well as for the choice of the right robot are given. This is followed by a product overview with the most important specific key figures.

Since the use of robots is only possible with the right tool, this will also be presented: Grippers, hands, as well as other accessories (sensors, image processing). A look at the aspect of occupational safety should show that a robot can be ready for operation within 10 minutes, but that some regulations must also be observed.

Notes on amortization calculations should serve as an economic decision-making aid. The already very short payback period can sometimes be further shortened by means of subsidies. Possible subsidies are outlined. The book concludes with an example of how the use of robots to bring processes that have been relocated abroad home can open up completely new business opportunities.

If the forecasts that predict an increase in the market volume from today's EUR 300 million to EUR 15 billion p.a. within eight years (source: Voith-Franka) are correct, then for small and medium-sized companies it means nothing more than "either MRK or dead". Today, many small and medium-sized companies name price and flexibility as their competitive advantages. However, companies with robots will be able to produce more cheaply and will be far more flexible. If required, the robots can work an extra shift over the weekend.

Those who doubt the forecasts are recommended to follow the sales development of the market leader Universal Robots. The parent company Terradyne has to publish regularly as a US-listed company. It is therefore well known that Universal Robots can generally increase sales by 20% compared to the same quarter of the previous year. Since Universal Robots is still losing market share, the market is growing even faster. This is not surprising either. Asian countries such as China, in particular, will have to invest massively in productivity growth as a result of the high wage increases there. For this reason, too, Kuka was acquired and a startup with Chinese shareholders, Yuanda, was established in Hanover, which should have good opportunities on both the German and Chinese markets. On the other hand, there are various Chinese Cobot manufacturers who have so far only produced for their domestic market. With Siasun, the laying of the foundation stone of a research centre in Magdeburg in the summer of 2019 has noticeably extended its feelers to Germany. Han´s Robotics is currently setting up a subsidiary.

Economically, the increased use of robots will be inevitable if today's level is to be maintained or even increased. For a lesson in economics states that nations can only develop positively if their productivity increases. This means that output must rise faster than wages. However, the latter have been rising significantly in the DACH region for some years now, so that they cannot be compensated only by common productivity improvements. Robots in particular therefore help to reduce unit labour costs. (It is no co-incidence that in the supposed low-wage countries - see above - such as China or Vietnam, Cobots are increasingly being used. In China, only about 4% of the population were middle class around the year, now the figure is around 70%.)

A brief look at the history of robotics

Since KUKA introduced the world's first industrial robot with good mobility thanks to 6 axes in 1973, the performance capabilities of classic robots have increased steadily. As a result of their high costs and lack of flexibility, however, the robots were reserved for larger companies with good or fully utilized series production. The rule of thumb, that the costs of programming and "around" exceed the pure (high) industrial robot price by a factor of 2, was as much a deterrent as the enormous expenditure of time. Quite a few robot projects in this area require a time span of more than one year, with fixed implementation often still six months. But when industrial robots start to work, it is often very impressive. Every reader is therefore recommended to visit one of the large automobile plants - our children were thrilled by their visit to the BMW plant in Munich, despite the two-hour walk. - Since not every company needed such large robots, over the course of time these robots were given smaller lightweight robots than siblings. However, the complexity of the furnishings was hardly reduced.

In 2008, the former startup Universal Robots founded the MRK (human-robot collaboration) class. Human-robot collaboration means the joint work of humans and robots. This means that the robots must not injure their human colleague, which is why various safety requirements must be taken into account. Instead of a fence as with the industrial robot, light cells, light barriers and much more were necessary. In order to work more flexibly than the big brothers, the Danish company had simplified the programming. Although the new robots sold well and Universal Robots grew impressively, the number of cobots sold worldwide p.a. (a different name for MRK) still reached just under 400,000 units. In 2008, the year the first UR robot was introduced, the company Rethink Robotics was founded in the USA. Their robots have a single screen head with two eyes and therefore look more human than the other Cobots. After the insolvency in 2018, even 150 million US-$ were not enough for the start-up financing, Rethink belongs in the meantime to the German Hahn Group.

In 2017, the Munich-based company Franka Emika entered the market. At the end of the year, Franka Emika received the German President's Future Prize endowed with 250,000 € for its only model, the "Panda". A "Panda" can only be compared to the smallest model of Universal Robots in terms of range and payload, but it is cheaper and easier to program. There are ready-to-use apps for various applications. Your own programming is also saved as apps and can then be repeated and bundled. Above all, however, the panda became famous with a video that was not meant to be imitated: he approaches his inventor with a knife and stops before he meets him. With higher speed he approaches a balloon in another video and stops in time. These aspects as well as the price of just over 10,000 € caused a sensation. The family-owned Voith Group, which had sold its Kuka shares at a high profit in 2016, acquired a stake in Franka in 2018. Together with the two companies that joined in 2018 (real start-ups) Kassow Robotics (Denmark) and Yuanda (German developers based in Hanover and headquartered in China), the low-cost suppliers Automata (UK) and mip-Robots (France), around 30 robot manufacturers are currently likely to offer well over 70 models.

It is foreseeable that the MRK (human-robot collaboration) will become increasingly cheaper and more efficient and will ultimately become a commodity product after the first boom in demand. Franka designed the production accordingly: Robots are to build the 15,000 pandas in Durach near Kempten, which is the medium-term goal, and will be supported by only 35 people with an 80% degree of automation. Franka also saves when it comes to purchasing: none of the parts comes from a European country, the supply chain starts in Turkey. The right grippers, sensors, optical systems and increasingly apps will be decisive for efficient use. Instead of the 15,000 pieces, Franka probably produced far less MRK in 2018, but the trend per Cobot remains positive. At the beginning of 2019, Franka was able to sell around 600 pandas per month. A publication by the University of Berkeley shows just how much the manufacturing costs of a Cobot depend on the number of copies produced. This company has developed the two-armed Blue robot, for which the following costs are estimated depending on the production volume:

Produciton Quantity (Arms)	10	50	250	1.500	10.000
Materials (BoM cost)	2.500	2.000	1.700	1.400	700
Labor	1.600	1.200	600	450	300
Testing	85	85	85	85	85
Material Markup	750	600	420	210	70
CM Operation	1.000	700	270	210	70
NRE`s (tooling, etc.)	17.200	3.440	688	144	30
Manufactured Cost per Arm	**23.135**	**8.025**	**3.763**	**2.499**	**1.255**

(Source: David V. Gealy et. al: Quasi-Direct Drive for Low-Cost Compliant Robotic Manipulation)

The market leader Universal Robots, whose market share is declining as a result of the new competition despite strongly increasing own production figures, celebrated the delivery of its 25,000 robot in 2018. A five-digit sales volume is already expected for 2019.

Annual growth of over 50% is forecast for the near future:

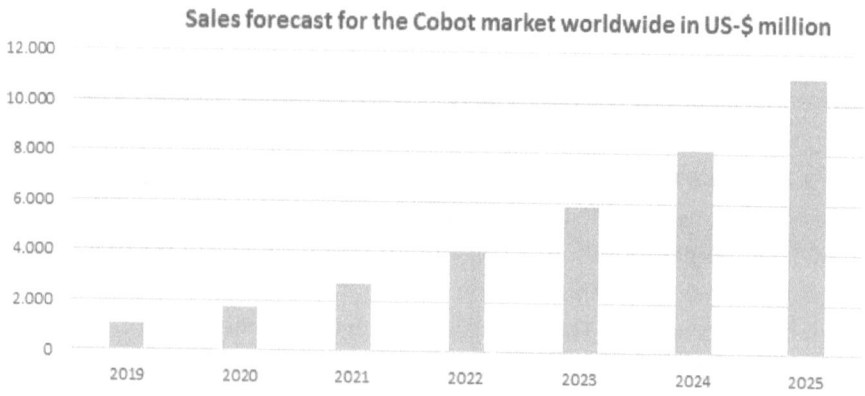

Sales forecast for the Cobot market worldwide in US-$ million

Source: MarketsandMarkets, Collaborative Robots Market –Global Forecast to 2025 lt. OnRobot

Decision-making aids for MRK use

Universal Robots, the MRK (human-robot collaboration) market leader, cites recurring manual tasks that require neither human skill nor critical thinking or creativity as ideal for robot use. This is particularly true for SMEs: the initial aim should be to automate support activities and gain experience. Step by step automation certainly makes more sense to accept the danger of too much complexity. The aspect of human dexterity is becoming less and less important, however, as the first grippers in the form of a human hand with movable fingers are available. Critical thinking is often associated with decisions to be made. A robot can only meet these with accessories - does the production piece have the right dimensions? More complex robot solutions make sense if the danger to a human being is reduced/excluded (operation of a press) or activities with a high repetition frequency are carried out.

The working speed should not exceed that of humans without further safety measures. The corresponding regulations ("speed limits") take into account the potentially touchable parts of the body as well as weights and surfaces.

A basic prerequisite for the selection of a new colleague, apart from his range, is his payload. This is to reduce the weight of the accessories (gripper, camera?). On the other hand, the same applies to robots as to humans: The force is lowest when the arm is outstretched. If this is angled, the efficiency is higher. Another criterion is the geometry of the object to be lifted. Each of us can lift 10 kg or more if they are modelled in the shape of a bucket, for example. We can hardly lift a 2 m long pole with one hand and hold it horizontally when its centre of gravity is at its end. This means that robots can carry far more than specified with suitable objects and the appropriate axle load. Nevertheless, in the event of an accident, the professional association will probably define the official payload as the limit of what is acceptable.

If the range of a robot is too small, the question arises whether different robots, perhaps also with different ranges, can be bundled or the layout of the production can be changed. By the way, the cooperation of two robots can reduce the acquisition costs if they are controlled by a controller. First Cobot manufacturers offer this option.

If no further accessories are to be used, more or less equally sized and equally placed parts to be handled by the robot are helpful. But as already mentioned, the optical solutions are becoming more and more versatile and in the meantime even bulk materials can be sorted. Nevertheless, robot beginners should initially do without the optics if they have simpler application options available. I.e. first concentration on the robot and then on the accessories. However, this should not be understood as a knock-out criterion if sensors or optics are absolutely necessary from the outset. Because it is better to do a little more work/complexity than not to use the chance of partial automation with robots. A compromise can also be a robot with integrated camera. His software will be able to recognize simpler parts easily, the effort for the company is then lower than with the training of additionally purchased software. However, the latter can recognize more.

If the robot has to handle pointed or sharp-edged objects at a higher speed, a human should only work at a safe distance. In addition to the robot, humans also pose a risk if they unintentionally come too close to the robot. Against this background, the very good (subscribe!) YouTube channel Next Robotics (german language) has differentiated the cooperation with humans into three categories:

1. Coexistence: Similar to today's industrial robots, humans and robots work separately and side by side. The robot can move at maximum speed, light barriers can be used instead of a cell (grid fence).
2. Cooperation: The robot works largely alone and only every few minutes with the human being, e.g. at a transfer point. The robot can then drive at its maximum speed - if no human being is nearby - and reduces this speed significantly if a human being approaches.
3. Collaboration: Both work together in the same place - here the strictest safety regulations apply, i.e. lower cycle times.

Despite the typically low payback period, it can be a risk limitation objective to invest as little as possible at the beginning of the MRK introduction. In this case, the question arises as to whether several grippers or only one should be purchased in addition to the robot. The reduction to only one gripper can limit the application possibilities. In addition to "pliers", suction cups, "drills" and much more can also be used as grippers. The American gripper startup RightHand Robotics will launch a gripper in 2019 that has a suction cup in the middle of its "hand" and three "fingers" around this middle. He can use it to suck articles out of a container and then hold them. The author is surprised that there is no gripper with two hands yet. These at a distance of a few centimeters could, for example, assemble (electric) parts and thus partially replace 2-arm robots such as the Yumi.

Of course, the price also plays a role. If you only want to use one or a few robots, you should be satisfied with a good amortization. If the amortization is guaranteed, in my opinion the last one must not be "tickled out" anymore. A Cobot of the manufacturers Universal Robots or Omron certainly costs more than that of some competitors, but it is also easier to learn and definitely of good quality. In the following also cheaper Cobots are introduced. These appear to be of particular interest for "mass applications" (an activity carried out by numerous MRKs) or for machine manufacturers who want to add a Cobot to their machines. Then the probably more complex programming or integration pays off.

An interesting case study is the experience report of the managing director of MS-Schramberg GmbH & Co KG. He points out that employee acceptance is increased if the robot initially performs unpleasant tasks for humans (please click on the picture or the link):

https://youtu.be/bUbTTVl73po

(Source: YouTube, Rethink Robotics)

By the way, not all robots can be used under the same conditions. This applies to the temperature range as well as to the dust and humidity concentration or the position. Some computers can work upside down (e.g. hanging from the ceiling - clever solution for use only at night; during the day the robot does not take up any space), others get "confused". This also applies to mobility. Not every model may be moved during operation. Robots like Franka's "Panda" would not understand this. Basically, this also applies here: There is (almost) always a solution - be it another Cobot or accessory. There is an upper limit of 50 degrees for the temperature. For higher temperatures, however, in my opinion special solutions are conceivable, e.g. a jacket made of an extremely insulating material.

Most MRIs have only one arm. By clever connection of two robots ("multi robot communication") almost all models can - at least theoretically - be converted to 2-arm MRK (human-robot collaboration), as this video shows (please click or link below):

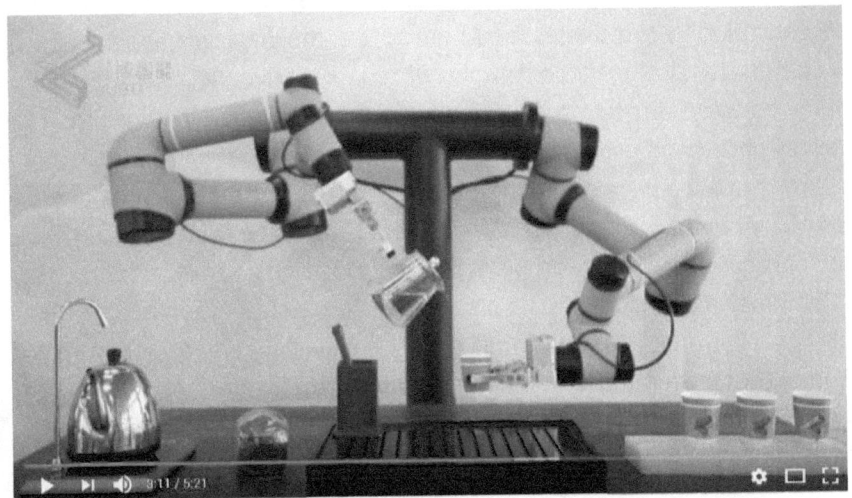

https://youtu.be/3y9N1l7ofYY?t=1m48s

(Source: YouTube, Aubo Robotics)

Alternatives to Cobots/ MRK

If the criteria mentioned do not fit or if fast working is desired, there are other interesting robots. The author, who works in an advisory capacity, is often commissioned by companies for whom special mechanical engineering as an automation option is too complex or too expensive. As a rule, there is almost always an alternative to a special machine when it comes to things such as packaging, processing or machine loading and the individual article is rather small and light. In addition to the classic and rather expensive industrial robots, which are also difficult to lift, so-called Delta or Scara robots are particularly suitable for packaging or other pick-and-place applications. Combined skilfully, these can also be used for an MRK (human-robot collaboration). The last mentioned robots are available for less than 30.000 € incl. good image recognition. The extreme performance of a Delta robot, for example, is demonstrated by this video, in which not only spoons are lifted, but also correctly positioned:

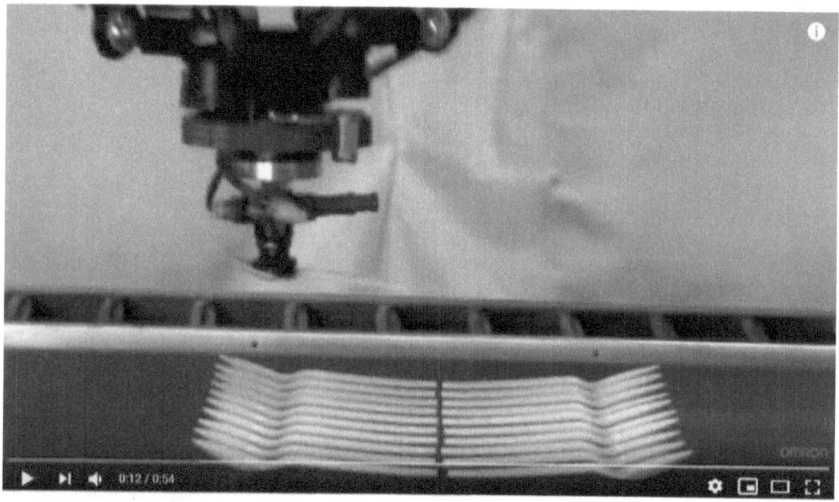

https://youtu.be/IKjq7Lecplw

(Quelle: YouTube, Omron Robotics and Safety Technologies, Inc.)

In addition, there are high-speed lightweight robots that can be very fast and do not pose a significant risk of injury due to their low weight. The following video shows a Motomini from Yaskawa, which is not a Cobot, but due to its low weight (7 kg) and its short range is hardly a danger, but for its applications (up to 0.5 kg payload) it is very fast, agile and precise:

https://youtu.be/PokMwjZSjTE

(Source: YouTube, Yaskawa America, Inc.)

Robot suppliers and their most important models

The following information refers only to the robot. The switch cabinets also weigh more than 10 kg, but are not considered here. While most attributes are self-explanatory, this does not apply to the "IP digits". They are therefore[1] explained below. Of the robots presented here the Motoman HC10 and the Franka Panda with 20 have the lowest protection, Fanuc and Stäubli with 67 the highest protection. Typical is the code number 54, which is also found in the models of the market leader Universal Robots, for example, which are often found under adverse conditions.

#	1st digit = touch	2nd digit = tightness
0	No protection against accidental contact	No water protection at all
1	Protection against foreign bodies > 50 mm	Protection against vertically falling water hits
2	Protection against foreign bodies > 12 mm	Protection against water hammering falling 15% to the vertical
3	Protection against foreign bodies > 2,5mm	Protection against spraying water falling 60% to the vertical
4	Protection against foreign bodies > 1,0mm	Protection against spray water
5	Complete protection against accidental contact, protection against dust deposits inside	Protected against water jets (from all directions)
6	Complete protection against accidental contact, protection against ingress of dust	Protected against ingress of water during temporary flooding
7		Protected against ingress of water during immersion
8		Protected against ingress of water during immersion for an indefinite period of time
9		Protected against ingress of water from any direction even at high pressure against the housing (high pressure / steam jet cleaner80-100 bar)

---

[1] http://www.reinmedical.com/de/technik/ip-schutzklassen.html

It is important to note that the controller and, if applicable, the teach panel (tablet) of the Cobot are sometimes assigned to a lower class. These parts can be protected, but if this is too cumbersome for you, you should pay attention to the same classes. The Teach-Panel can be protected against the consequences of a fall by a rubber armoured protective cover, regardless of the environmental conditions.

And: The controller is not just an appendage, but its performance can limit complex programming in the worst case.

ABB

ABB introduced its YuMi IRB 1400 several years ago. Therefore he has no app yet, but as one of the few two arms. Due to its short reach (55 cm) and Payload (0.5 kg per arm, after deduction of the grippers only about 0.2 kg), it is predestined for the production of small parts such as electronics (extreme repeat accuracy!), Lego bricks or watches. Its price of about 40.000 € is quite high and is rarely surpassed by the robots presented in this book. These aspects make it a niche product with USP for larger companies, e.g. with production lines.

Model YuMi	2-arm	1-arm
Number of arms	2	1
Degrees of freedom	7	7
Range mm	559	559
Payload kg	0,5	0,5
Payload by hands kg	0,2	0,2
Own weight kg	38,0	9,5
Repeat accuracy mm	0,02	0,02
Ambient temperature range	5-40	5-40
IP	30	30

Due to its short range, it is preferred in electronics or for fun cooking, as the video shows:

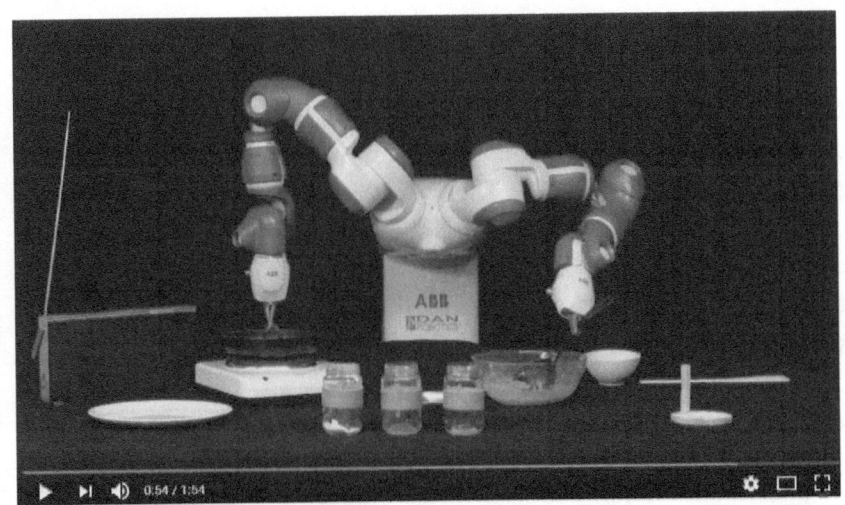

https://youtu.be/U7XL9_0dSJs

(Source: Youtube, DanRobotics A/S)

For the target group of this book it seems to be less interesting. This probably also applies to the one-armed brother "Single-arm YuMi" introduced in 2018.

Acutronic

Acutronic Robotics was only founded in 2016, has its operational headquarters in Spain and has included the Japanese Sony group among its investors since 2017. At the end of 2018, the company presented a new MRK concept and launched its first model, "MARA". The special feature of MARA is that each of the seven Degrees of freedom can be operated independently thanks to H-ROS, so that individual Degrees of freedom can be dismantled. Instead of seven axes, MARA also works with only two axes, which reduces costs. Like other Cobot providers, Acutronic points to the low programming effort thanks to artificial intelligence. Acutronic uses - at least in theory - particularly advantageous AI tools, so that, for example, the desired can also be simulated. At the same time, MARA can be networked so well with other suitable hardware and the programs can probably also be exchanged. An axis can also be exchanged for third-party equipment, e.g. a camera.

While these options make MARA interesting for complex applications, the author sees a further advantage in the low price (from 15,000 €) and the good dirt resistance. With IP 54, MARA is superior to the German PANDA with a similar range and carrying capacity. While the PANDA can be used above all in the fields of electronics and precision mechanics, MARA also has no problem with industries such as food or CNC.

model	MARA
Number of arms	1
Degrees of freedom	6
Range mm	656
Payload kg	3,0
Own weight kg	21,0
Repeat accuracy mm	0,10
Ambient temperature range	0-50
IP	54
Special:	modular design

Two videos are recommended for understanding:

The first video introduces the topic in a more theoretical way:

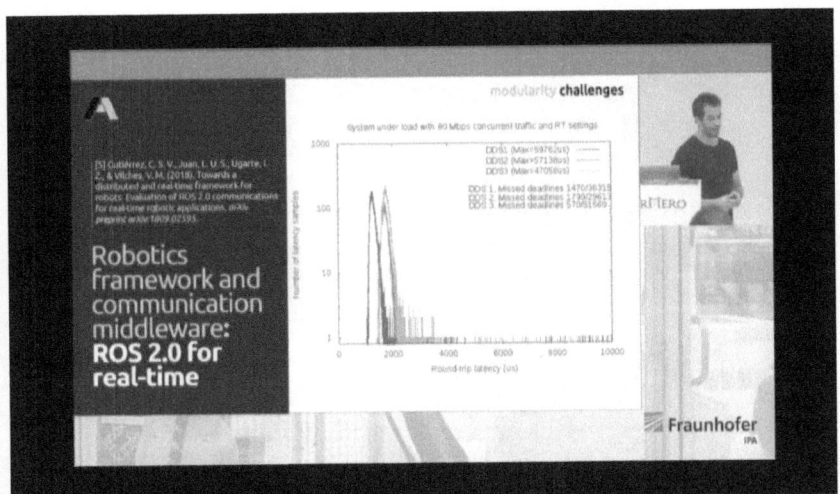

https://youtu.be/otZi43tdWvo

(Source: YouTube, Acutronic Robotics)

While in the video above the robot practically does not appear, it can be viewed here:

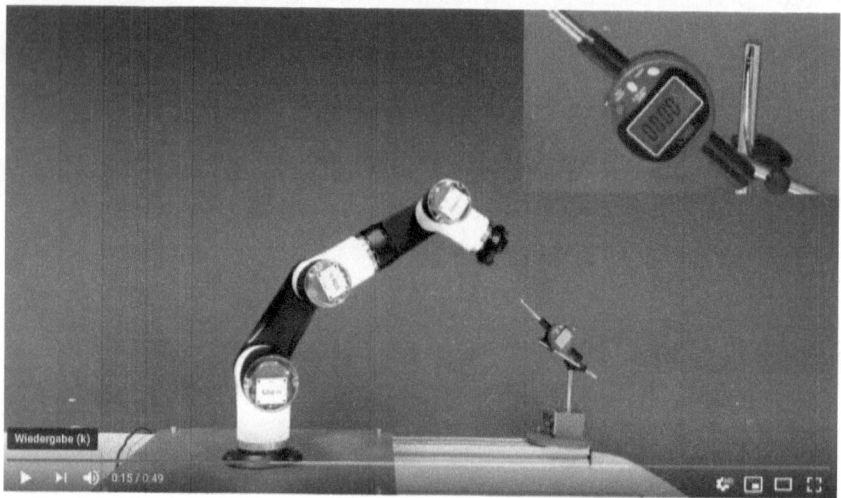

https://youtu.be/40JzCmWwk3o

(Source: YouTube, Acutronic Robotics)

Aubo

This US-American robot manufacturer is currently trying to open up the European market and also offers an autonomously moving lower part. The video shows the use of several Aubos in a Chinese production facility and thus confirms that robots also pay for themselves in China. With a price of only about 15.000 € this is not surprising. Its accessories should be compatible with Universal Robots.

model	i-5
Number of arms	1
Degrees of freedom	6
Range mm	924
Payload kg	5,0
Own weight kg	24,0
Repeat accuracy mm	0,05
Ambient temperature range	0-45
IP	54

The linked video shows that MRK use is also economical in China despite the lower wage costs there. The corresponding company uses 60 (!) Aubos in the CNC area.

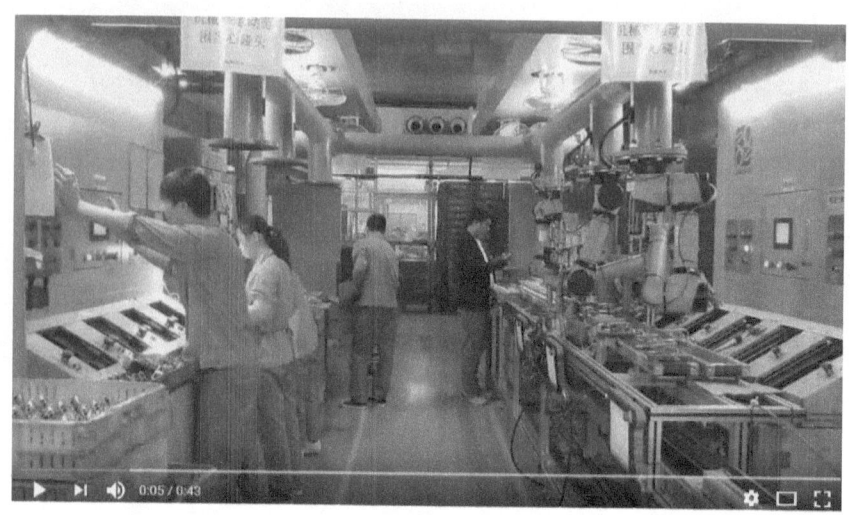

https://youtu.be/xLTtPrciej0

(Source: YouTube, Aubo Robotics USA)

Automata

The British startup can be described as the price breaker among the suppliers. At the Hannover Messe 2019 it presented its only model with "EVA". The equivalent price of just under €6,000 is likely to be difficult to undercut at present. The ABB investment of $7.4 million can be seen as a accolade for the company's founders. As ABB is particularly active in the Chinese market, which is calling for low-cost solutions, the main sales market will soon be China.

model	EVA-2
Number of arms	1
Degrees of freedom	6
Range mm	600
Payload kg	1,3
Own weight kg	9,5
Repeat accuracy mm	0,50
Ambient temperature range	5-40
IP	20

The Cobot reminds of the Panda of Franka from its performance data and its intended use cases (e.g. test of electronic articles), which is however still more mobile with 7 Degrees of freedom.

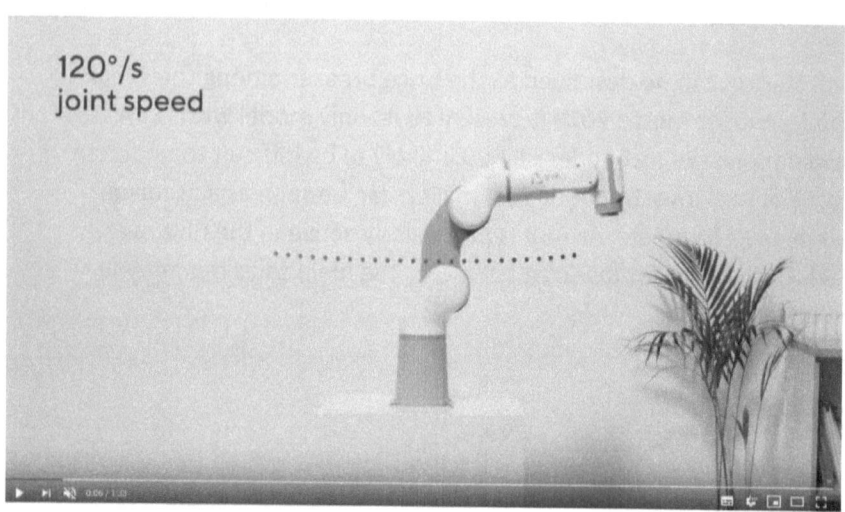

120°/s
joint speed

https://youtu.be/Df_XHu2HRrQ

(Source: YouTube, Automata)

One thing is for sure: The purchase price can be expected to include support such as Universal Robots or Omron. The useful life could be another weakness. However, with the investor ABB in the background, there should also be an awareness of quality. The grabs offered also come from Schunk.

Bosch

As a technology group, Bosch - like FESTO - feels a connection to robotics. Consequently, its stationary and mobile robots are not intended for individual workplaces (where other manufacturers would have an advantage in terms of price/performance ratio), but as an integral part of entire production lines. Bosch uses Kuka robots as required.

model	APAS assistant inline	APAS KUKA
Number of arms	1	1
Degrees of freedom	6	6
Range mm	911	1.100
Payload kg	5,5	10,0
Own weight kg	35,0	60,0
Repeat accuracy mm	0,03	0,01
Ambient temperature range	n.a.	n.a.
IP	54	54

The special feature of the APAS devices is their very high repeatability, which is not required for most applications.

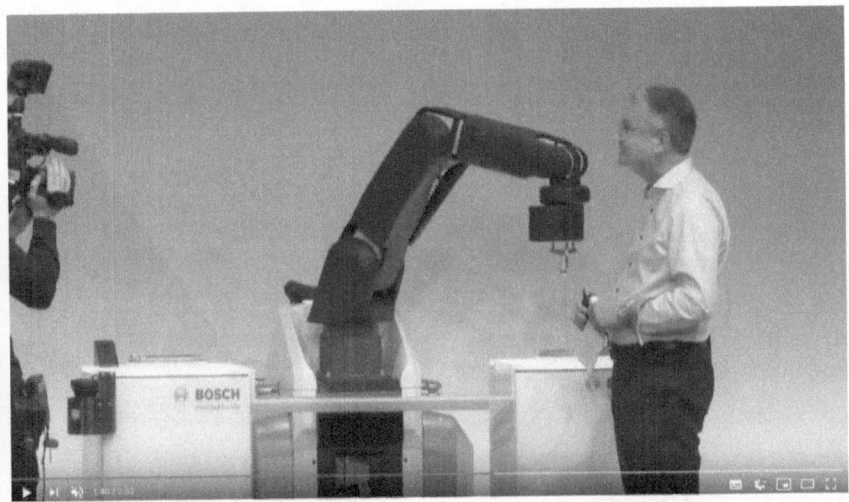

https://youtu.be/stFYQaHULLk

(Source: YouTube, Bosch APAS)

Doosan

The South Korean group is known as an excavator manufacturer. The fact that he has entered the robot market shows what potential is generally seen in MRK. Doosan intends to enter the European market at the end of 2018. One USP of its robot series is the unusual reach of up to 1.70 m and the announced payload of 15 kg. There are also versions for potentially explosive environments.

model	M0609	M1509	M1013	M0617	N.N.
Number of arms	1	1	1	1	1
Degrees of freedom	6	6	6	6	
Range mm	900	1.500	1.300	1.700	1.700
Payload kg	6,0	15,0	10,0	6,0	15,0
Own weight kg	27,0	32,0	33,0	34,0	
Repeat accuracy mm	0,10	0,10	0,10	0,10	
Ambient temperature range	5-45	5-45	5-45	5-45	
IP	54	54	54	54	
Special:		Anti static available			
tablet programming	x	x	x	x	x

The Doosan is used in various industries in Asia. This video shows the range quite well:

https://youtu.be/gX2D16RWRek

(Source: YouTube, Don Don)

Doosan has its own sales organization in Germany.

Elephant Robotics

The Chinese company presented its "Catbot" at the Hannover Messe 2019, which costs only about 5,000 €. However, the Cobot is not (yet) available in Germany and probably not in Europe. Since the documentation of the "Catbot" is available in English in contrast to its siblings listed below, it can be assumed that the "Catbot" will also be offered here - even via the new Alibaba logistic centers.

The interesting thing about the "Catbot", apart from its high-quality optical impression, is the promised high repeat accuracy and options such as voice control. Especially voice control, if it works, can be the key for cheap robots like the "Catbot" to give craftsmen or hobbyists a third hand. As usual with the super cheap MRK, no tablet is included in the scope of delivery. Since in the meantime everyone has a private tablet, this is not a major problem, so that nothing stands in the way of programming, if you can.

model	catbot	Panda 3	Panda 5
Number of arms	1	1	1
Degrees of freedom	6	6	6
Range mm	60	50	810
Payload kg	3,0	3,0	5,0
Own weight kg	18,0	17,0	24,5
Repeat accuracy mm	0,05	0,05	0,05
Ambient temperature range	0-50	0-50	0-50
IP	42	42	42

At the low price, the Catbot can be used for simple activities, as the video shows (units of measurement are not visible, though).

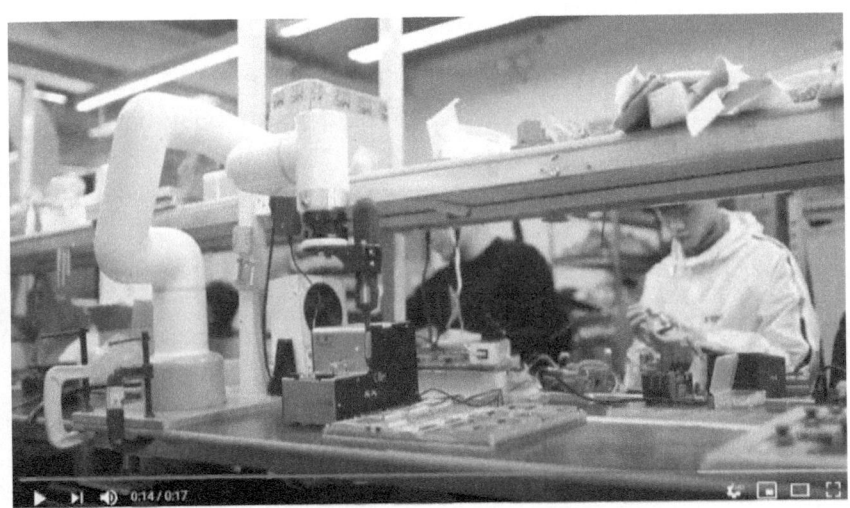

https://youtu.be/uOKjqMWDygM

(Source: YouTube, Elephant Robotics)

Elephant Robotics has no distribution in Germany or Europe. EU certifications are therefore also likely to be lacking.

ESI (Engineering Services Inc)

Driven by the extreme growth potential in the Cobot market, the Canadian company ESI entered the Cobot market in 2018. His first two robots have an a.o. high repeatability and an equally unusual force (15 kg) and reach of up to 132 cm. However, their flexibility is limited by the unusually high weight. With weights of 100 kg incl. controller, it is not easy for even two employees to clear away the Cobot.

Whether the robots will also be distributed in Germany or Europe is not yet foreseeable. In the course of further development, they should be able to drive autonomously. Camera solutions are already available today. With the automotive supplier Magna, an interesting reference customer has already been won.

model	C-7	C-15
Number of arms	1	1
Degrees of freedom	6	6
Range mm	900	1.323
Payload kg	7,0	15,0
Own weight kg	50,0	100,0
Repeat accuracy mm	0,05	0,05
Ambient temperature range	0-40	0-40
IP	54	54

As can be seen in the video, ESI constantly scans its surroundings for obstacles and stops working when, for example, a person is nearby.

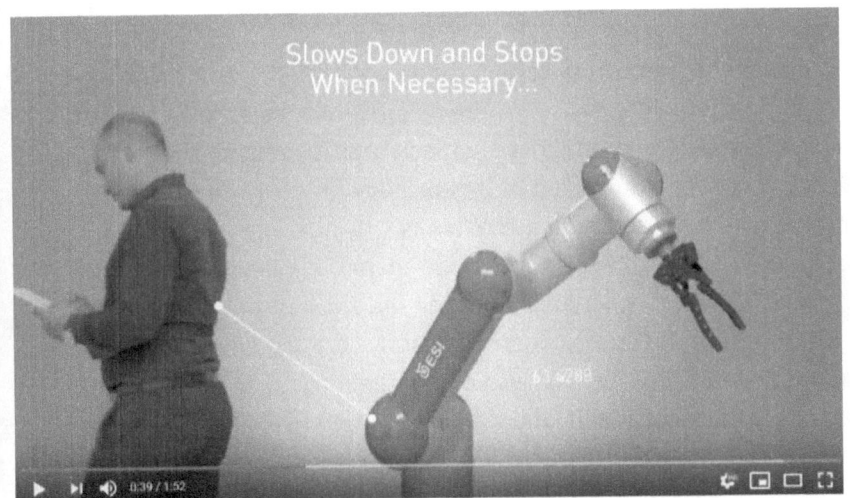

https://youtu.be/mhnUwyP_MCM

(Source: YouTube, Engineering Services Inc.)

Fanuc

The Japanese robot manufacturer offers the strongest MRK. Its lifting power of 35 kg should be uninteresting for the majority of readers, especially as it costs a high five-digit amount. Possible special applications such as changing car tires to rims are actually out of the question, as they can hardly pay for themselves with only seasonal use. What is striking is the "clumsiness" of all models and their high weight, but also - thanks to a high protection class - the arbitrariness of their location. Even the smallest of the robots listed below costs about €40,000 - perhaps hardly competitive - and the largest about €80,000. But Fanuc is also one of the most experienced providers with a high level of quality and support awareness.

model	CR-4iA	CR-7iA	CR-15iA	CR-35iA
Number of arms	1	1	1	1
Degrees of freedom	6	6	6	6
Range mm	550	717	1.441	1.813
Payload kg	4,0	7,0	15,0	35,0
Own weight kg	48,0	53,0	255,0	990,0
Repeat accuracy mm	0,10	0,10	0,10	0,10
Ambient temperature range	0-45	0-45	0-45	0-45
IP	67	67	67	54/67

A video shows the enormous power of the largest model, but also makes the high weight of almost 1 to understandable:

https://youtu.be/U7XL9_0dSJs

(Source: YouTube, FANUC Europe)

If the 35 kg payload is not sufficient, the Italian company Comau offers sensitive Scara robots (2 axes), which can lift up to 120 kg.

F&P Personal Robotics

The F&P-Cobot comes from Switzerland, with which almost everything would be said. Its price of about 25,000 € is perhaps high in view of its Payload of 5 kg, but for a Swiss citizen the price premium is manageable.

model	P-Rob 2R
Number of arms	1
Degrees of freedom	6
Range mm	775
Payload kg	3,0
Own weight kg	20,0
Repeat accuracy mm	0,10
Ambient temperature range	
IP	54

In the following video an F&P packs, for hygienic reasons well packed, hanging (not everyone can) and mobile (also not everyone can) yoghurt pots with the help of numerous suction cups.

https://youtu.be/b_M5-XDjmqE

(Source: YouTube, F&P Personal Robotics)

It seems as if the company is now specializing in gastronomy robots and those for care (including massage).

Franka

The "Panda" from the Munich base company Franka, which was launched on the market in 2017, has significantly increased the attention paid to MRK in the media and made the Cobots virtually suitable for the mass market thanks to its extremely low price and extremely simple programming including distribution in the cloud. As the first MRK, it is also CE-compliant. The disadvantages are its lower Payload, the moderate range and the problems to be expected as a result of the low protection class in less favourable environments. With its 7 degrees of freedom, the panda is particularly mobile.

model	panda
Number of arms	1
Degrees of freedom	7
Range mm	859
Payload kg	3,0
Own weight kg	18,0
Repeat accuracy mm	0,10
Ambient temperature range	15-25
IP	20

The Panda can lift 3 kg just like the UR3 and should therefore be a real alternative to this. A robot with a payload of 10 kg was announced for 2020. Probably he will then also have IP 54.

A video shows how easy it is to set up and program the panda:

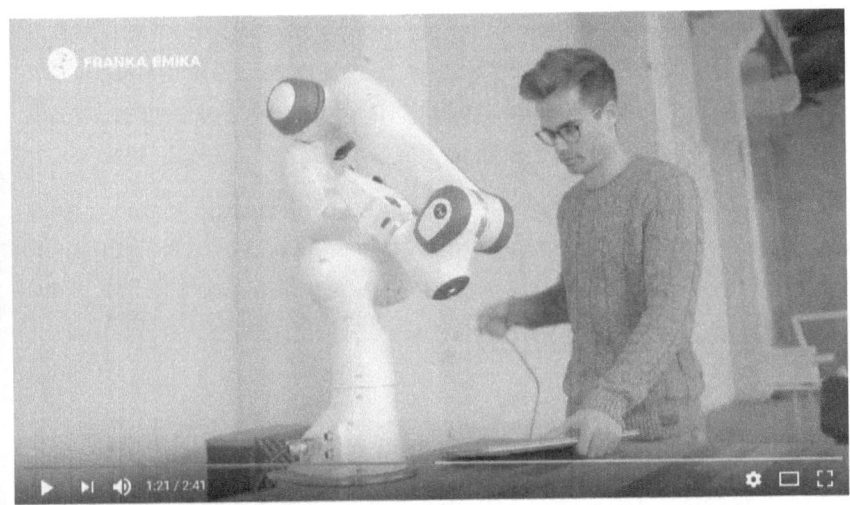

https://youtu.be/bXo68UFNyhk

(Source: YouTube, Franka Emika)

The Panda World is a platform where programming and solutions created by other users can be purchased. At the beginning of October 2019, the program library already counted 100 apps, i.e. solutions for different applications. Since sales are now expected to be around 1,000 units per month, there are already numerous applications. The Eco-System around the Panda has probably exceeded the critical size long ago and offers solutions even for CNC, whereby the low payload can limit the use. Against the background of the already large eco-system, the risk of buying a panda is limited. This is all the more true in view of the fact that the startup with the family-owned Voith Group has a financially strong shareholder. Voith is also responsible for sales, so that interested parties can also access the corresponding website.

https://voith.com/robotics-en/index.html

Moreover, hardly any other company is as well connected as the co-founder of the company, Prof. Dr.-Ing. Sami Haddadin. Haddadin has been director of the newly founded Munich School of Robotics and Machine Intelligence at the TU Munich since mid-2018.

Another plus point at Franka is likely to be the integrators enthusiastic about the panda. On the Internet (YouTube or Twitter) you can actually see pointless solutions (e.g. opening a beverage bottle) that are programmed only by the person who has fun.

The panda is not only particularly sensitive, but also likes to experiment: If it cannot put something on in the first attempt, for example, it tries it further by increasing its radius. But in this video everything works out on the first try:

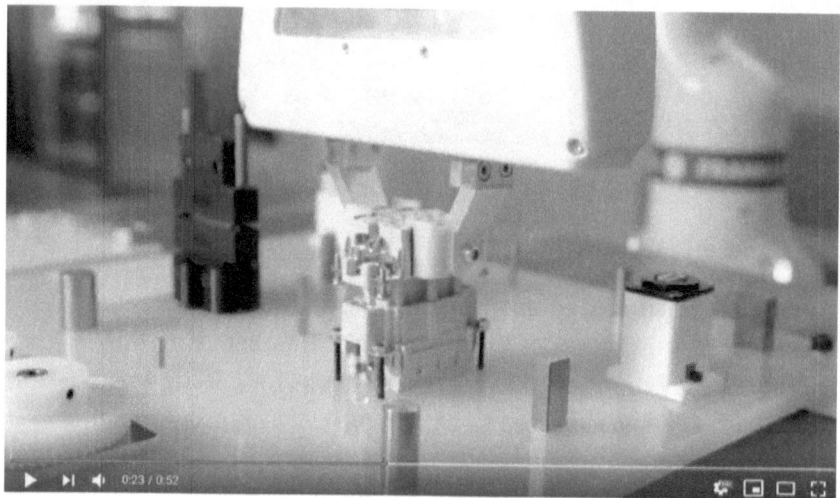

https://youtu.be/Fy8dnS45YyA

(Source: YouTube, Franka Emika)

The author waits with excitement for the second Franka robot. This does not necessarily have to be characterised by a higher payload and longer range. The idea of a care robot and thus leaving the classic Cobot genre is also conceivable.

HAN´s Robots

The Chinese HAN´s group now also serves Germany from Spain. (A small curiosity: The domain "hansrobot.eu" was derived from "HAN´s-Robots".) The website of the German, Munich-based distributor lists numerous use cases and highlights the integration with Kuka, ABB and Fanuc. In March 2019, a German GmbH was also founded in Reutlingen, but its managing director lives in Switzerland. It appears that the German branch will not commence operations until mid-2020.

Currently, three different models are offered, which roughly correspond to the technical data of Universal Robots. As a price query on Alibaba showed, HAN´s should not be much cheaper than Universal Robots. Technologically, they make a good impression - including possible voice control - so far probably unique in the Cobot sector. HAN´s and Acutronic are partially compatible.

model	E3	E5	E10
Number of arms	1	1	1
Degrees of freedom	6	6	6
Range mm	590	800	1.000
Payload kg	3,0	5,0	10,0
Own weight kg	17,0	23,0	40,0
Repeat accuracy mm	0,05	0,05	0,05
Ambient temperature range	0-50	0-50	0-50
IP	54	54	54

The video shows one HAN´s soldering:

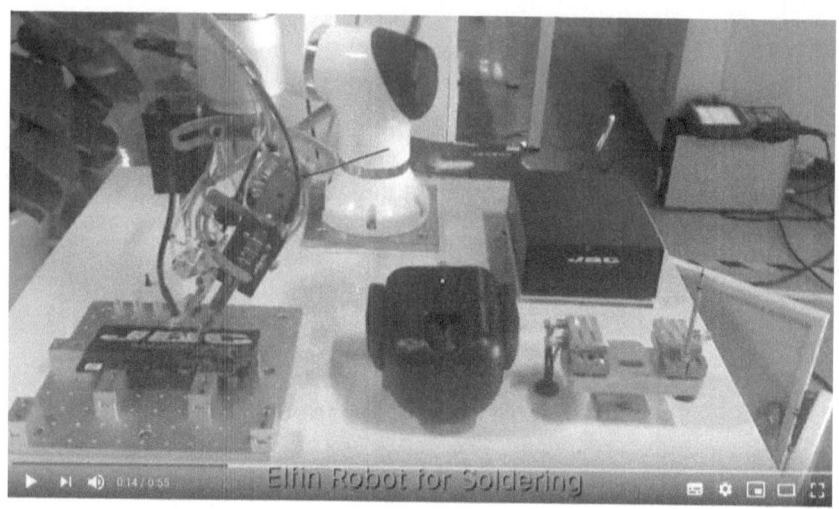

Elfin Robot for Soldering

https://youtu.be/CYSve-tVKoo

(Source: YouTube, Han´s Robot)

Hanwha

The Hanwha conglomerate is headquartered in South Korea and made its first appearance in Germany with the takeover of the solar cell manufacturer QCells. South Korea is known for its high technological competence. The Hanwha is therefore not a cheap robot, but a highly solid, very easily programmable Cobot, which has already won the iF Design Award 2017 in Germany.

model	HCR-5
Number of arms	1
Degrees of freedom	6
Range mm	915
Payload kg	5,0
Own weight kg	20,0
Repeat accuracy mm	0,10
Ambient temperature range	0-50
IP	54

That Hanwha has a special claim is shown by the video with the quite complex programming.

± 0.1mm repeatability

0:58 / 2:31

https://youtu.be/L_uyBH7K4s

45

The video suggests a vision: The current collegial use can hardly be programmed, but it can be done with speech recognition ("Alexa" & Co.). For example: "Turn the frame by 20 degrees").

A former Rethink integrator sells and installs Hanwha robots in Germany.

Jaka

In April 2019, the Chinese start-up Jaka received US $ 15 million in Series B financing, a total of US $ 39 million. In view of these investments, it can be assumed that Jaka will also try to gain a foothold in Europe, at least in the medium term. The equipment of the Cobots seems to be suitable for this.

model	Zu3	Zu7	Zu12
Number of arms	1	1	1
Degrees of freedom	6	6	6
Range mm	498	796	1.300
Payload kg	3,0	7,0	12,0
Own weight kg	12,0	21,0	31,0
Repeat accuracy mm	0,03	0,03	0,03
Ambient temperature range	0-50	0-50	0-50
IP	54	54	54

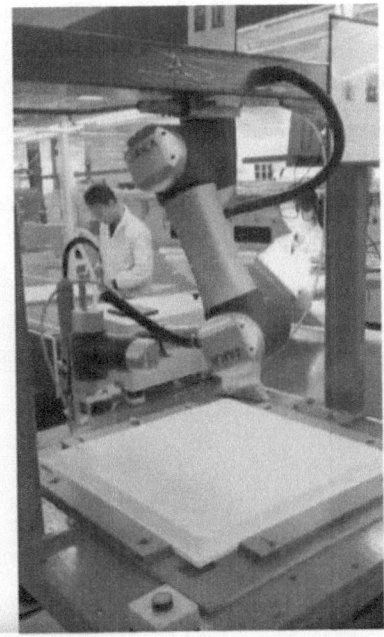

https://youtu.be/fY4S-PPnYE4

(Source: YouTube, Louis Lee)

Kassow Robots

The Danish startup with the Russian name presented itself for the first time at Automatica 2018. The robots seemed a bit simple, especially in comparison to the Hanwha, but have an interesting performance spectrum:

model	KR810	KR1205	KR1805
Number of arms	1	1	1
Degrees of freedom	7	7	7
Range mm	850	1.200	1.800
Payload kg	10,0	5,0	5,0
Own weight kg	23,5	25,0	45,0
Repeat accuracy mm	0,10	0,10	0,10
Ambient temperature range	0-50	0-50	0-50
IP	54	54	54

In recent months Kassow has been able to gain several distributors in European countries and the USA, including Germany. Kassow can also be an option because the founder of the same name was one of the founders of Universal Robots. In other words, he knows both the technology and the market and is therefore likely to have easier access to investors if things get "stuck". The video shows a clear advantage of the Kassow-Cobots: Their strength, here the robot lifts two 2.5 litre bottles and thus 5 kg.

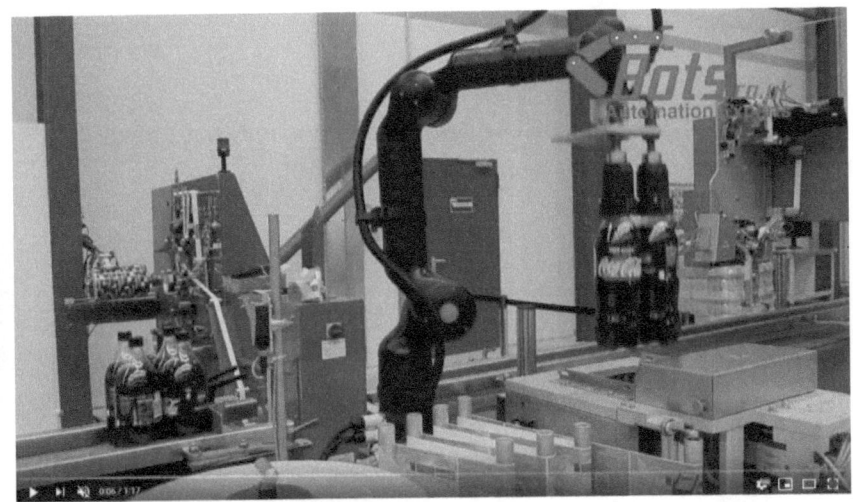

(Source: YouTube, Bots Automation)

Kassow is determinedly building up the international sales network and is researching voice control for his robots. Typical commands like "open hand" etc. already work very well. In addition, an integrator has already merged a Kassow with a MIR robot to form a hybrid robot.

Kawada Industries

The mobile Nextage, equipped with two arms and optics, may have been ahead of its time for years (it was launched around 2010).

model	nextage
Number of arms	2
Degrees of freedom	6
Range mm	577
Payload kg	1,5
Own weight kg	29,0
Repeat accuracy mm	
Ambient temperature range	

For special applications, as can be seen in the video (laboratory), he can of course pay for himself - price: 60,000 euros.

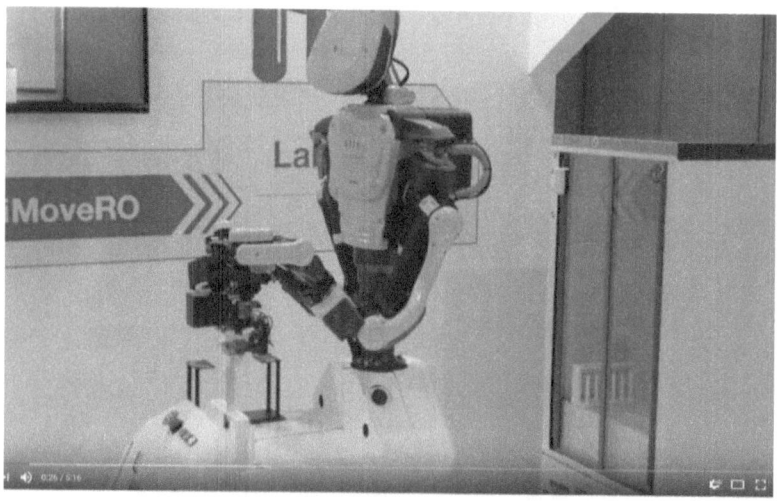

https://youtu.be/a8ly8L_rrMM

(Source: YouTube, Industrial Robots)

Kuka

The top dog among the industrial robot manufacturers has recognized the trend and pressure of the MRK offers a compact Cobot, the iisy, which will appear in 2019 and is supposed to be the delayed answer to the Panda of Franka. Franka initially wanted to work with Kuka, but then realised that the business models were contradictory. Kuka has so far generated high revenues from programming, which Franka does not envisage as its own revenue model. As a result, the former Kuka major shareholder joined Franka. The iisy should cost about 20,000 € and would therefore correspond to the price expectations / budget of the target group of this book. The iiwa costs about 60.000 €. Its use therefore pays off above all in multishift operation, but not only with occasional use.

It should not go unmentioned that Kuka belongs to a Chinese company and that German companies have been avoiding Kuka in part since the takeover for fear that data etc. might get to China. A supplier for a high-tech company could thus have problems with his customer if he showed him his production including Kuka robots. Pure speculation, but worth mentioning.

With its short range, the iisy is designed for the electronics and medical industries. The very expensive iiwa has already been available for over six years - so programming is not very easy yet.

model	iiwa 7 R800	iiwa 14 R820	iisy (from 2019)
Number of arms	1	1	1
Degrees of freedom	7	7	6
Range mm	800	820	600
Payload kg	7,0	14,0	3,0
Own weight kg	23,9	30,0	18,8
Repeat accuracy mm	0,15	0,15	
Ambient temperature range	5-45	5-45	
IP	54	54	

For the iiwa there are autonomously driving coasters, as can be seen in the video:

https://youtu.be/9WNE3JAcO6U

(Source: YouTube, Kuka Robots & Automation)

Mabi

Another swiss manufacturer, who offers beside the robots similar to Kuka also mobile undercarriages. High-end devices that can be particularly fast without MRK throttling and above all compete with Kuka. Probably unbeatable: Can be used from all angles. However, sensitivity appears to be still limited at present. The market says that Mabi will make a new start in 2020 with new Cobots. Sales to date have been very modest.

model	Speedy 6	Speedy 12
Number of arms	1	1
Degrees of freedom	6	6
Range mm	800	1.250
Payload kg	6,0	12,0
Own weight kg	28,0	35,0
Repeat accuracy mm	0,10	0,10
Ambient temperature range	0-55	0-55
IP	54	54

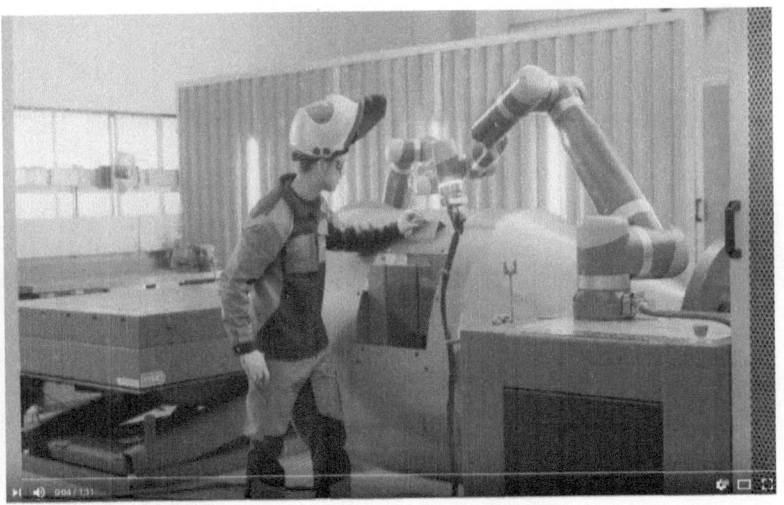

https://youtu.be/mbVIJdKTiQ8

(Source: YouTube, Mabi AG Robotic)

A small excursion: In the video you can see a technician in a white coat with a futuristic glasses attachment. These are simple data glasses, such as Garmin already offers for 300 € for cyclists including networking to their bike computer:

https://youtu.be/ppjoDXQ0VEE

(Source: YouTube, DC Rainmaker)

MIP-robotics

This French start-up pursues an original, but probably not very popular approach on the market. It replaces two axes with one vertically movable axis, which can move 188 mm in height or depth. Presumably for this reason the own weight as well as the price (about 8.000 €) is below average low. Whether MIP regards its products as fully-fledged is doubtful in view of the choice of name for the models.

model	Junior 200	Junior 300
Number of arms	1	1
Degrees of freedom	4	4
Range mm	40	60
Payload kg	3,0	3,0
Own weight kg	13,0	13,5
Repeat accuracy mm	0,10	0,10
Ambient temperature range		
IP	20	20

Programming via ROS is an advantage for the expert. A vision system for the optical recognition of articles is available as an accessory.

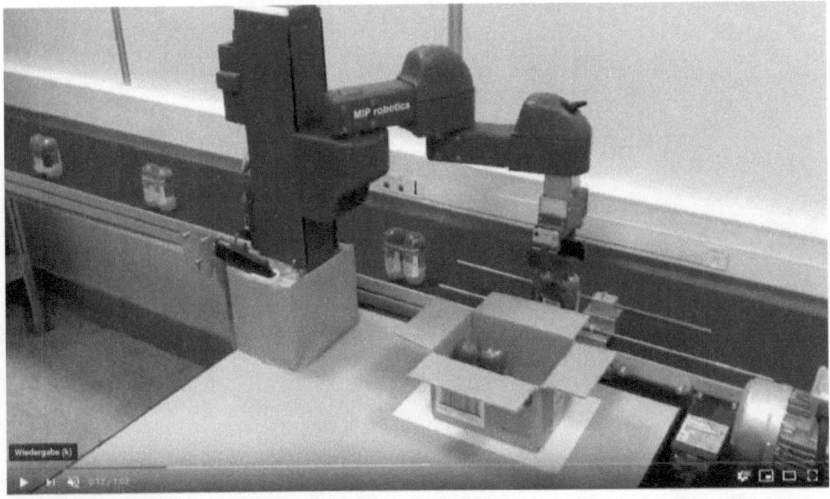

https://youtu.be/d6Nr2vPkof8

(Source: YouTube, MIP robotics)

Motoman

The Japanese company Yaskawa is known in Germany as Motoman. His two very similar robots are versatile (ceiling, wall), but the price of around 40,000 € for the target group of the book seems quite high. Interested parties should check whether the protection class is sufficient.

model	HC 10DT	HC 10
Number of arms	1	1
Degrees of freedom	6	10
Range mm	1.200	1.200
Payload kg	9,0	9,0
Payload by hands kg		
Own weight kg	48,0	48,0
Repeat accuracy mm	0,10	0,10
Ambient temperature range	0-55	0-55
IP	20	20

The protection class is low.

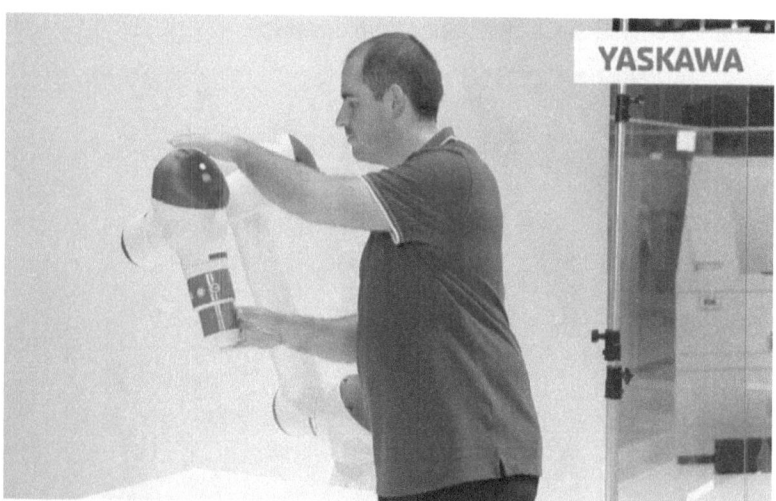

https://youtu.be/YGjRIEQ1xoM

(Source: YouTube, Yaskawa Europa GmbH)

With the DT (Direct Teach) model variant, the robot can be trained directly on the arm by pressing the appropriate buttons.

In mid-2019, the French company E-Cobot integrated a Yaskawa on its mobile robot "Husky".

Nachi

The Japanese company presented its first two Cobots at the end of 2017. They are easy to teach by means of teaching and have a touch-sensitive outer skin. During operation the connection with data glasses is possible.

model	CZ5	CZ10
Number of arms	1	1
Degrees of freedom	6	6
Range mm	1.300	1.300
Payload kg	5,0	10,0
Own weight kg	35,0	51,0
Repeat accuracy mm	0,10	0,10
Ambient temperature range	0-45	0-45
IP	54	54

In addition to the long range, a special feature is the ability to work hanging from the ceiling.

https://youtu.be/c9F3_3yMEOk

(Source: YouTube, Nachi Robots)

Omron

The Japanese Omron Group is regarded as the automation specialist with the world's largest number of articles. For some years now he has consistently turned to robotics - also with the help of larger acquisitions (adept) or cooperations. Omron is a premium supplier. This is already evident optically: no other Cobots look higher-quality - perhaps a criterion for use in public spaces. There are few competing products with this mixture of payload and range:

model	TM5-700	TM5-900	TM12	TM14
Number of arms	1	1	1	1
Degrees of freedom	6	6	6	6
Range mm	700	900	1.300	1.100
Payload kg	6,0	4,0	12,0	14,0
Own weight kg	22,1	22,6	33,3	32,6
Temperature range	0-50	0-50	0-50	0-50
IP	54	54	54	54

The prices start at about 25.000 €. The Omron Cobots have a built-in camera (5 MB) with clever image recognition. Not only can it easily recognize parts, but it can also read barcodes and helps the MRK to orient itself well at all times. The "landmark" concept developed in-house is used for this purpose. Landmarks are stickers. The robot is taught where to find a sticker. When he sees him again, he already knows where he is.

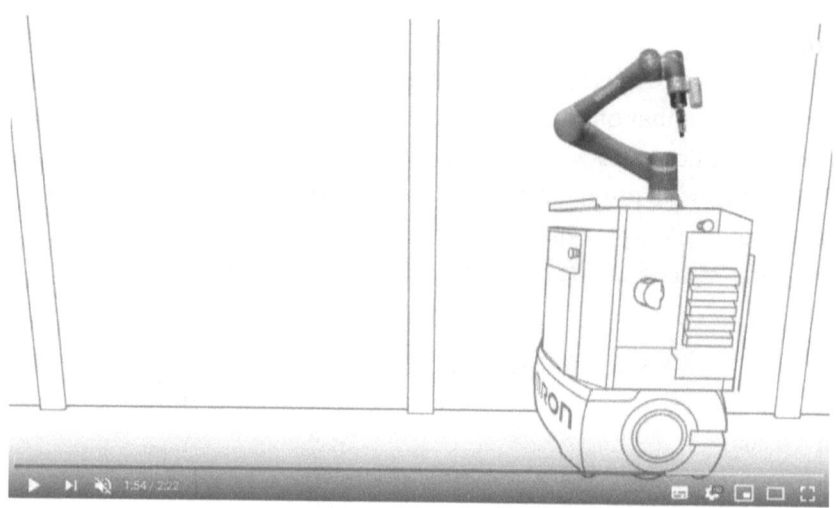

https://youtu.be/xRQIftQ0Tvw

(Source: YouTube, Omron Industrial Automation EMEA)

In autumn 2019, Omron plans to have its Cobots married to its AIV (Autonomous Intelligent Vehicles). The AIVs are autonomously moving transport robots that - depending on the model - can transport 60 to 130 kg and, at 30,000 €, cost about as much as a Cobot today. This makes it possible for a Cobot to operate several machines or to drive to the warehouse on its own to fetch something (it can read barcodes). Fleet management enables the accident-free and parallel use of up to 100 AIVs.

Of interest for belt production: In the automotive industry in particular, the workpiece to be machined is moving, people are going along and screwing etc.. Omron allows the robot to work in parallel.

Another interesting detail at Omron is the presence of other Delta and Scara robots that can be connected to the Cobot. This means that comprehensive (mobile) complete solutions are possible from his hand.

Like Universal Robots, Omron has an extensive sales network in Germany with its own regional showrooms.

The company is a partner in the "Opdra" joint project, which is presented in a separate chapter in this book and in which machine screens are read for the first time. An Omron-Cobot then makes any necessary corrections.

Productive Robotics

The American company was apparently able to sell so many OB7 models in its home market that it was able to finance the development of two further models from current income. Productive Robotics is one of the few suppliers whose robots have seven instead of six axes and are therefore even more mobile. The company prides itself on offering the easiest to use Cobots. The price of about $20,000 for the OB7 is about the usual level.

model	OB7	OB7-Max8	OB7-Max12
Number of arms	1	1	1
Degrees of freedom	7	7	7
Range mm	1.000	1.700	1.300
Payload kg	5	8	12
Own weight kg	24		
Repeat accuracy mm	0,1		
Ambient temperature range	n.a.		
IP	54		

A powerful vision system for easy learning of objects is offered as an accessory. If the company enters the German market, it should have good prospects.

https://youtu.be/2szyhWL3BWI

(Source: YouTube, Productive Robotics)

Rethink Robotics

With the Baxter, the American MRK pioneer initially offered a two-arm robot. The Sawyer followed this model in 2015. Both MRK have in common that they have a tablet as a kind of head. And since this screen also shows a face in work mode, Rethink offers the most human Cobot - at least optically. The consciously human character of the robots is also explained by the choice of name: Sawyer stands out from the usual type designations of other manufacturers. The quality deficiencies existing under the US leadership are to be remedied in the meantime under the new German leadership (see below), but must be taken into account when purchasing used equipment.

model	Sawyer
Number of arms	1
Degrees of freedom	7
Range mm	1.260
Payload kg	4
Own weight kg	19
Repeat accuracy mm	0,1
Ambient temperature range	0-40
IP	54

Rethink has its own interesting gripper family called "ClickSmart", which enables a change of activity within a very short time. Also practical: there is a mobile and adjustable frame. The above-average price of these Cobots can be explained by the extensive equipment with cameras.

The photo shows the application in a rather improvised US filling plant (no German Engineering). But if an entrepreneur who accepts such a system spends money on a robot, then it must be worth it.

(Source: Rethink Robots)

The video shows that the Baxter, which had been adjusted in the meantime, was able to replace a human being, whereby its low load-bearing capacity must be taken into account.

https://youtu.be/fCML42boO8c

(Source: YouTube, Rethink Robotics)

The Rethink robots certainly belong to the premium products. Despite investments of 150 million US-$, the company had to file for bankruptcy in September 2018. In October 2018, the German Hahn Group, which has been active in robotics for 30 years, took over the patents and know-how. Hahn intends to further develop the robots and at the same time make the Rethink software available to other MRK manufacturers as part of licensing models. In future, at least four Cobot series will come from Germany (Rethink, Kuka, Franka and Yuanda). Since mid-December, Hahn has been eagerly promoting Rethink on social media channels like Linkedin. Hahn is backed by the Ruhrkohle Foundation.

At the beginning of 2019, the CEO of the world market leader Universal Robots, von Hollen, commented on Rethink: "One of the best competitors we ever had was Rethink" (Source Boston Clobe, 01.01.2019). Probably for this reason Universal Robots took over 20 former Rethink employees. Rethink tries to hire more employees, but finds it difficult because of the peripheral location in the Hunsrück region.

The company boasts that its Cobots are extremely easy to train.

Rethink has also been offering its robots on loan since mid-2019. The possibility of renting robots instead of buying them should be an option both for the cautious and for seasonal operators.

Siasun

Siasun is the name of a fast-growing Chinese robot manufacturer that is looking for sales partners worldwide and offers an interesting range of products. This includes, among other things, the seventh degree of freedom as well as an integrated vision system with 1.3 million pixels but rather weak.

model	XCR20-1100	SRC 5	GCR-1100
Number of arms	1	1	1
Degrees of freedom	6	7	6
Range mm	1.100	800	1.100
Payload kg	20	5	20
Own weight kg	50	33,8	50
Repeat accuracy mm	0,1	0,02	0,05
Ambient temperature range	0-45	0-45	0-45
IP	54	54	54

Especially interesting: There are autonomous coasters, so Siasun offers hybrid Cobots (gripping & driving). We deliberately did not link to a "Cobot-Video", but one that represents the cooperation with BASF. Should Siasun ever enter the German market, it is likely to be a strong competitor to the incumbents.

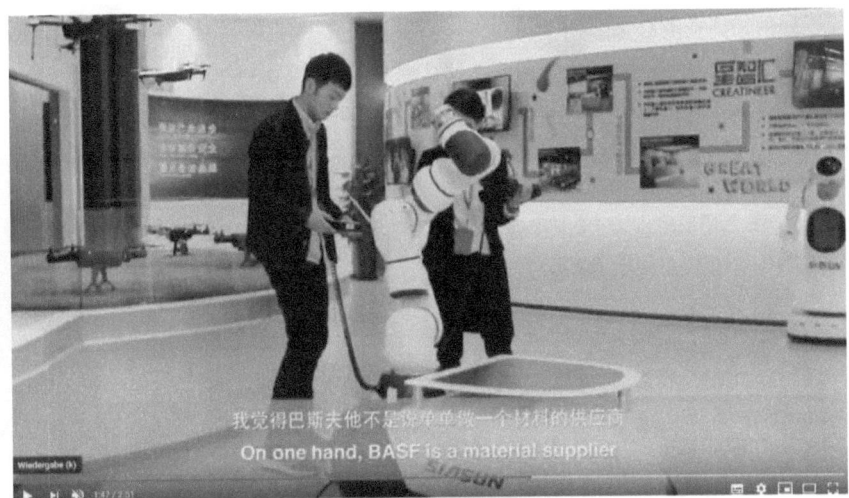

https://youtu.be/Gf0CDqb7CPo

(Source: YouTube, MassDevice)

With the completion of the construction of a research and development centre in Magdeburg, which began in the summer of 2019, the European certifications should also be available, so that Siasun can be expected to enter the German market by then at the latest.

Stäubli

Stäubli's two robots are certainly among the best quality on the market. The massive layout of the Cobots is relativized, considering that under certain conditions the Payload is almost twice as high (9 and 5 kg, respectively). The robots can also be mounted on the wall or ceiling. To the author's knowledge, both MRK are to be programmed like industrial robots, i.e. less suitable for the target group here.

model	TX2-60	TX2-60
Number of arms	1	1
Degrees of freedom	6	6
Range mm	670	920
Payload kg	4,5	3,7
Own weight kg	51,4	52,5
Repeat accuracy mm	0,1	0,1
Ambient temperature range	5-40	5-40
IP	65/67	65/67

The protection classes are a.o. high.

https://youtu.be/SYGO2NpXEmc

For the price of almost 40.000 € it requires accordingly fastidious work. Probably trump with their speed, which they can only live out, if they do not work collaboratively.

Universal Robots

The world's largest Cobot manufacturer seemed to have come under some technological pressure since 2017. Its three models had proven themselves, are considered to be very robust and relatively easy to program, but the new MRKs of the competitors were partly even easier to handle and also have apps (e.g. Franka Panda). At Automatica 2018, however, Universal Robots achieved a breakthrough. The existing robots are now offered as "e-series". According to advertising, the first own programming should be ready for operation after only one hour. Here 14 ready-to-use apps help. Allegedly there would already have been an amortization after only 34 days (4-shift operation?). The Teach Panel and the new intuitive user interface should contribute to this.

In addition to a force-torque sensor, which ensures greater sensitivity and thus a wider range of applications, the focus is on the faster and simpler implementation of robotic applications already mentioned, as well as improvements in occupational safety (17 safety functions such as programmable stopping time or braking distance are new, up to now there have been 15 functions). The e-version can also process more data, which is particularly helpful for controls based on optical information.

The prices of the previous models are between 15.000 € (UR3) and about 30.000 € (UR 10).

(Source: YouTube, Advanced Motion Systems, Inc.)

In September the previous trio was extended by the UR16e. With a reach of 90 cm, it can lift a remarkable 16 kg and thus compete with comparable robots of the brands Doosan, Kassow or Omron. Especially in the field of packaging and CNC, the UR16e should score well. Its power consumption of 350 kw is noticeably higher than that of its little brother, the UR3e, which manages with 125 kw.

model	UR3e	UR5e	UR10e	UR16e
Number of arms	1	1	1	1
Degrees of freedom	6	6	6	6
Range mm	500	850	1.300	900
Payload kg	3	5	10	16
Own weight kg	11	18,4	28,9	33,5
Repeat accuracy mm	0,1	0,1	0,1	0,1
Temperature range	0-50	0-50	0-50	0-50
IP	54	54	54	54

As representatives of IP class 54, Universal Robots can also withstand unfavourable conditions such as high humidity or oily air. For this reason, they frequently feed machining centres, for example. In addition, new grippers from third-party suppliers are generally offered initially or exclusively UR-compliant. With installations in nearly 40,000 production environments, UR has the most experience of any vendor.

The "Universal Robots Plus", which can be viewed on the Internet, includes interesting accessories of very different types and from different manufacturers, all of which have been certified by UR to be compatible with the Cobots.

In addition, there are numerous complete solutions (e.g. welding stands or machine assembly) that work with a robot from Universal Robots. Put simply, third-party products are always UR-compatible first.

It remains to be seen whether the mobile robots of MIR, a sister company (both bought by Terradyne), will be married to the Cobots of UR in such a way that the robots will be driven and powered by MIR robots and both robots will be controlled together. UR would then be the third supplier, along with Omron and Siasun, of hybrid robots, i.e. robots that can both drive and manually operate. In the area of hybrid robots, there are also integrator-built mergers, such as Kassow with MIR.

University of Berkeley

Although currently only presented and correctly available only in 2020 (advance orders are already possible at a higher price today) and although the whole will probably land in a spin-off (as Franka Emika or Yuanda), "Blue" is already presented. Because with Blue it concerns one of the rare 2-arm robots, in addition to that to a prognosticated fight price around approximately 5,000 €. Its 7 degrees of freedom as well as the integrated AI show what is already possible for little money. It is currently being advertised as a "household robot", perhaps because it would then not be able to withstand the high industrial burden.

After all, the reach of 70 cm with a payload of 2 or 4 kg (depending on whether only one arm or both are used) and the low dead weight of only 8.7 kg make it highly interesting. In general, the author thinks that there is still great market potential in the area of 2-arm robots. Whether it's for making up (cables) or for simple manual activities up to shrimp spools (the German shrimps are still transported to Morocco by truck and there up to 4,000 women are desking).

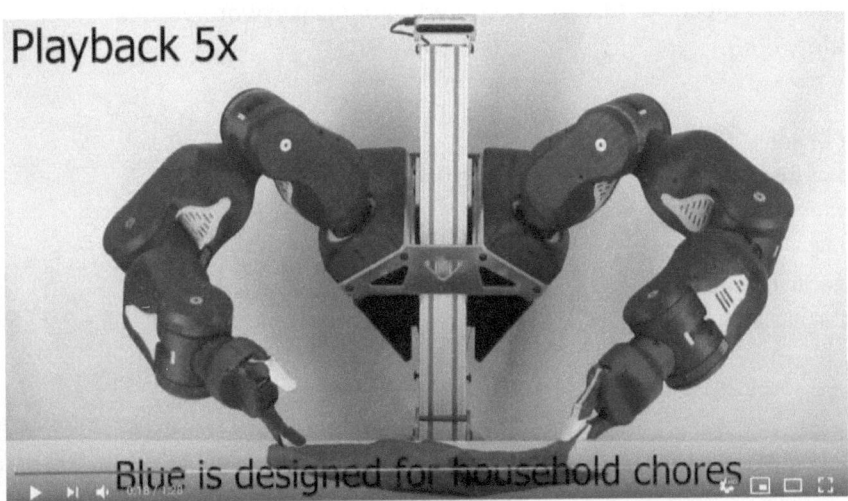

Playback 5x

Blue is designed for household chores

https://youtu.be/KZ88hPgrZzs

(Source: YouTube, UC Berkeley)

I can also imagine Blue as a kitchen helper.

yuanda

Yuanda is a start-up company based in Hanover, Germany, which emerged from the chair of mechatronics there and is financed by the Chinese building materials group of the same name, which wants to diversify. The fact that the robot was developed to market maturity within one year by only 20 people shows the low entry barrier to the market, provided that know-how is available. This can be taken as evidence for the thesis that MRK can only become cheaper - or will include more and more performance at the same price - similar to what we are used to from computers or mobile phones. The Yuanda is likely to be one of the cheapest robots and will be built primarily in China and launched on the German market in 2019. A second model has already been announced. A smaller production facility is planned for Germany.

The special thing about the Yuanda robots will be their modularity. The photo shows that the areas to be touched have been highlighted in green. Each robot has an integrated camera and software (object recognition, gripper control) adapted to it. An interesting approach is to control the environment by means of further integrated cameras (keyword occupational safety etc.).

model	M 6
Number of arms	1
Degrees of freedom	6
Range mm	1.000
Payload kg	7
Repeat accuracy mm	0,1
Ambient temperature range	0-50
IP	54

https://youtu.be/izYHHmAcRyY

(Source: YouTube, Kollmorgen)

In the future it will be possible to use a Microsoft Hololens for teaching. I.e. the "programmer" can teach the robot by gestures without having to touch it, which makes learning even easier. The Hololens, however, still costs about 5,000 € today.

At a VR Business Club event, the author was able to try out the Hololens and was convinced of their benefits. This video (a small excursus) shows the possibilities of the Hololens, e.g. remote maintenance (less qualified colleague, e.g. overseas, is instructed by the headquarters). The photo (video) shows how the Hololens takes up the perspective of the external colleague. The expert in the head office sees everything and can mark the relevant places and give hints to the external colleague via included telephone.

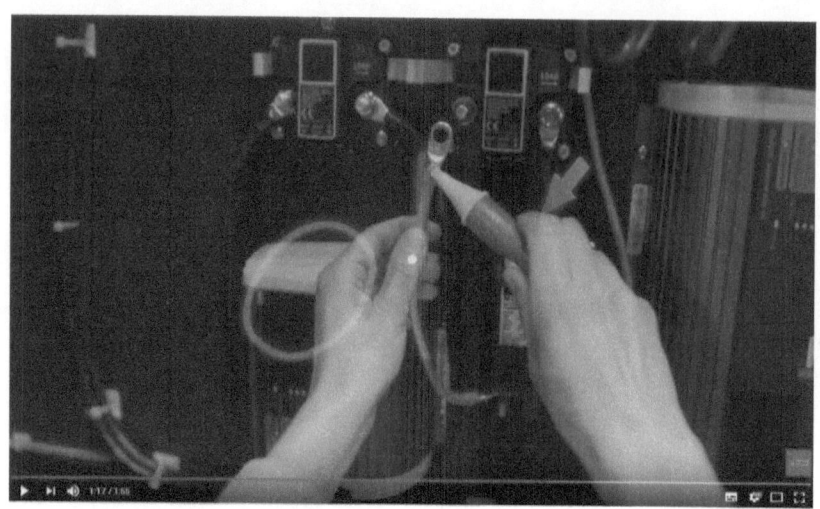

https://youtu.be/UpmolMrf5HQ

(Source: YouTube, Microsoft HoloLens)

Other manufacturers

In addition to the listed manufacturers, there are others whose robots are not yet available from us to buy or which do not seem to fit at all. Possibly in the foreseeable future also available in DACH could be the BRABO of the Indian company Tata Motors, also known in our country.

Suppliers like the Belarusian company Rozum Robotics seem to specialize in industry solutions. Rozum, for example, equips rows and rows of cafés in the Ukraine. A normal Cobot works there as a barista.

The manufacturer Neuromeka, based in South Korea, should probably meet the German quality requirements and is also cheap, but is not (yet) offered here.

For the daring who do not necessarily want to rely on the market leader: At the end of 2018, the Chinese Internet platform Alibaba announced the construction of a logistics centre in Liège with an area of 220,000 square metres. Further centres are to follow, thus opening up the European market. Products from China should then reach their European customers within three days. This is interesting, as Alibaba offers numerous robots on his Chinese website and it can be assumed that these Cobots will be available quickly and without problems - parallel to the existing distributors.

In total, there are said to be 120 different models.

Accessories

For many applications, a robot as supplied is not sufficient. It usually starts with the holder for the robot. It should be fixed on a very stable and fixed base, especially if it can lift a lot and has a greater range or moves faster.

Gripper

The hand of the robot is naturally decisive for success. The object to be lifted must not be lost or "tangled". The selection of suitable grippers is becoming ever larger and there are now even grippers that are modelled on the human hand - but still cost around € 5,000.

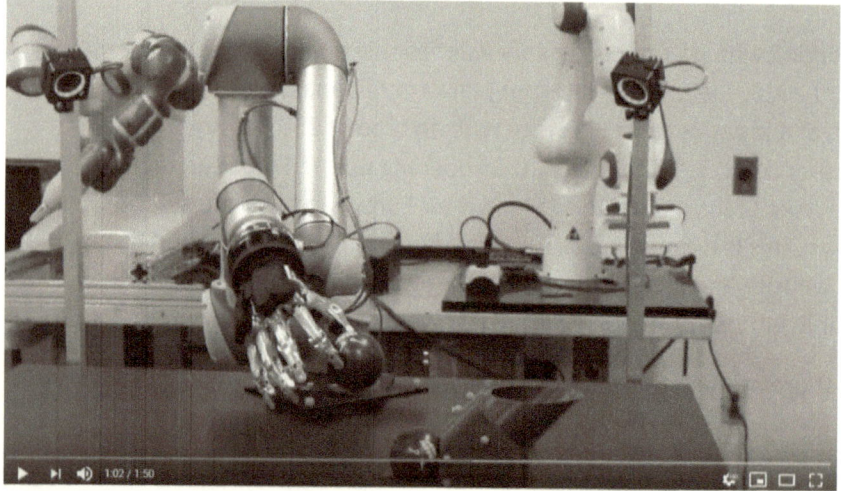

https://youtu.be/tGn9gpmPrms

(Source: You Tube, Energid Technologies)

Particularly sensitive grippers are a must for special applications. Grippers can not only grip, but can also suck (e.g. to lift plates or handle a wide variety of geometries in pick and place applications) or be magnetic. If a gripper does not have to move (open-close), a self-constructed gripper from the 3D printer can also be an ideal solution. The video shows at the beginning how such a gripper is printed. With this gripper a bottle of beer can be opened afterwards and the "gripper" can lift a beer bottle and a glass at the same time and pour the beer into the glass. A smart idea that you have to come up with first:

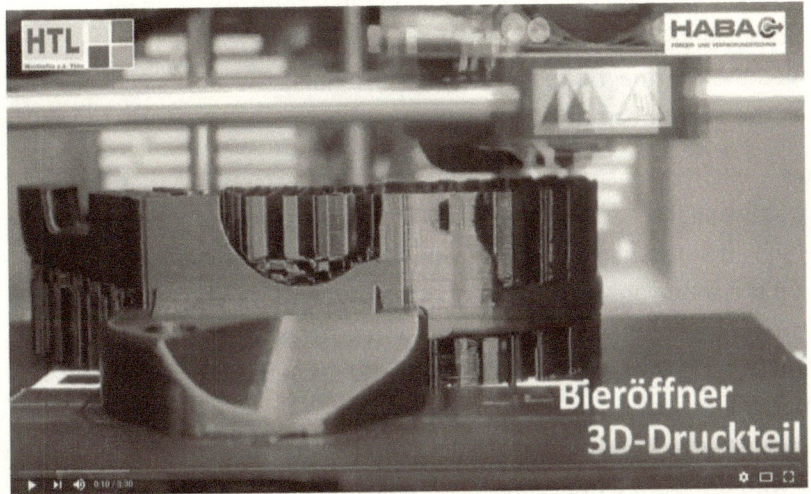

Grippers are offered both by the MRK manufacturers in numerous variants and by special equipment manufacturers. The German world market leader Schunk, such as the Danish company OnRobot, Robotiq from Canada or Weiss (also from Germany) should be mentioned here above all. With Robotiq, for example, a hand camera can be attached to the "joint" of the robot.

On its website, Schunk has extensive presentations of both standard products and customer-specific solutions. In 2017, Schunk received the prestigious Hermes Award, one of the world's leading technology awards. To see what a seemingly simple "on-and-off gripper" can do, look best for the "Schunk JL1". The Schunk MRK grippers are equipped with environment sensors and collision protection, among other things, in the interests of occupational safety. Schunk has recently started offering its MRK grippers under the new brand name "Co-act".

The gripper manufacturer Weiss Robotics has particularly sensitive, but also extremely fast grippers in its range. Gripping cycles of up to 500 pieces/minute are feasible. The gripping force can be safely and easily limited. One model is equipped with sensorless gripping force control so that even brittle, fragile or yielding parts can be moved without any problems. Weiss also developed the "Permagrip technology", which holds the workpiece in place even in the event of a power interruption.

OnRobot has a smaller selection, but grippers that are so sensitive that they stop closing at the slightest resistance. Particularly in the area of vacuum grippers, which lift something by suction, special solutions can be useful depending on the shape and weight distribution of the part to be lifted. When selecting the gripper, its payload, the spread stroke options, its closing speed and its weight must always be taken into account. This reduces the payload of the robot arm. In view of the potential risk of contact and thus injury, the selection of rounded grippers can be recommended. If different parts with different dimensions have to be handled, a double gripper can be a solution. With the acquisition of Purple Robotics in August 2018, OnRobot now also offers a variable vacuum gripper that can lift up to 10 kg without compressed air and can also have a second hand.

By the way, simple grippers are already available from 30 € in the online shop, but it should be checked here whether you are also safe. Plug & Play grippers are naturally more expensive. Plug & Play grippers are not compatible with every robot manufacturer. Here Universal Robots is favoured as market leader.

Here a suitably "Gecko" baptized and award-winning gripper from OnRobot lifts a glass plate. In contrast to typical vacuum grippers, materials with holes (there is no vacuum) or porous parts can also be lifted. No compressed air is required, so operation is easier and no supply lines are necessary. The required force is determined and requested by "Gecko" itself:

(Source: YouTube, OnRobot)

Vacuum grippers, unlike the Grecko, require compressed air, especially if they have to lift heavy plates etc.. They can also lift flexible bags. The company Piab has developed suction cups that can be attached individually, i.e. in a suitable way:

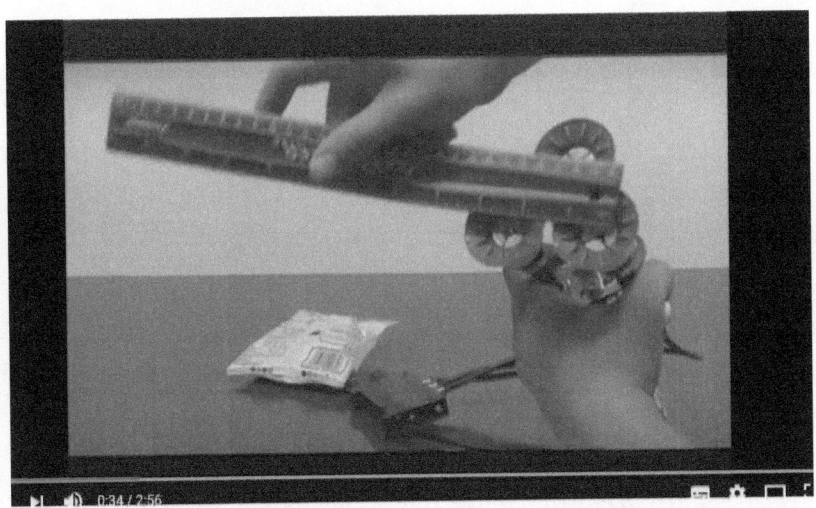

(Source: YouTube, PIAB Vacuum Technique)

Larger panels can be lifted with grippers positioned far apart from each other.

The American startup RightHand Robotics offers a mixture of finger gripper and suction cup. This company received 21 million US dollars (coincidentally as much as Robotiq received) in December as part of a further financing round and now wants to address e-commerce retailers in particular. The already functioning concept is as simple as it is convincing: If articles are "stuffed" closely next to each other in a carton, even the human hand has problems removing an article. Because there is no space between the individual articles for our fingers to grasp them. RigthHand Robotics has therefore integrated a suction cup in the middle of the gripper. Thanks to optical recognition and artificial intelligence, he can recognize and analyze the situation. Typically, the gripper first sucks the article up until its three fingers can grip it. Problem solved! (Articles of almost any shape can be handled.)

https://youtu.be/jjubkPajDTU

(Source: YouTube, RightHand Robotics)

Anyone who has ever had bad experiences is no longer bound to be so today. Specifically, the author was involved as a consultant in a project where biscuits had to be sucked in. The production manager was more than sceptical, because the solution he had got to know years ago was quickly ruined by crumbs, so that suction was no longer possible. But we also found a solution for this.

No robot project should fail because of the missing suitable gripper!

Tools

Instead of the classic grippers, the first companies offer tools that can be connected directly to the robot arm and thus further increase its range of applications. The MRK screwdriving system presented by Weber Schraubautomaten independently screws screws with a screw head size of 4 to 14 mm with a torque of 1 to 10 and a speed of 485-800 rpm (depending on torque). The system is available in two versions: One without its own protection concept and thus for integration into systems with an existing protection concept ("SEV-L") and a sensitive system with its own protection concept ("SEV-C"). With a weight of less than 5 kg, the tool can also be guided by MRK with a medium payload.

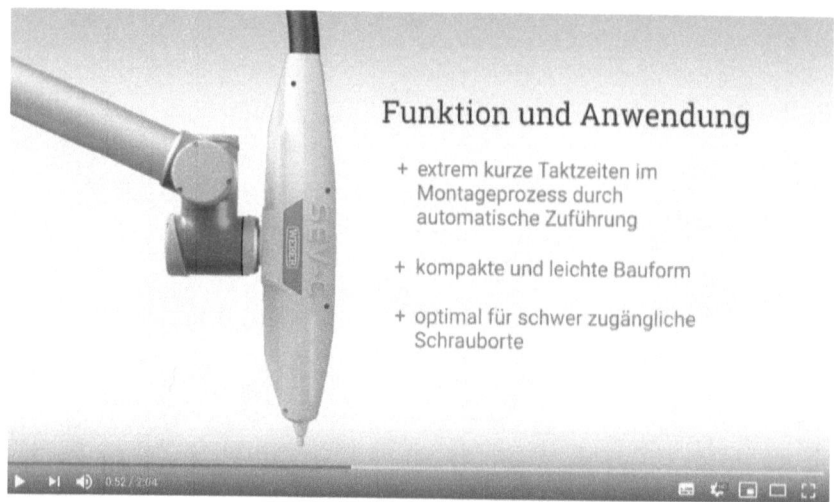

https://youtu.be/gtpiOtxmO54

(Source: YouTube, WEBER Screwdrivers)

The company Stöger Automation offers stronger and therefore heavier automatic screwdrivers.

The development of further tools for a wide variety of applications is to be expected. The potential manufacturers come from industries far removed from MRK and may first have to become aware of the MRK potential.

optics

The time when optical solutions were very expensive is over. Today, there are standard solutions - either as part of the Cobot (e.g. Yuanda or Rethink Robotics), as accessories for the gripper or as plug-and-play accessories from optical specialists.

Essentially, the following tasks are to be supported:

- Object recognition (incl. completeness check - are there 10 parts?, correct position adjustment etc.)
- Measurement (quality assurance)
- Recognize colors and contrasts
- Identification (code recognition, because an object recognition recognizes the object, but what if it is produced again and again and should be recognized again in terms of traceability?)
- positioning
- Surface inspection (quality assurance, is there delamination etc.?)

The video of the Munich startup Roboception shows how easy object recognition can be.

https://youtu.be/-leZbn1aA8Q

(Source: YouTube, Roboception)

Very interesting appears the possibility of automatic quality control by means of a laser scan by Robotiq. The video section on the left shows the gripper, the section on the right the measurement results as they appear on the screen:

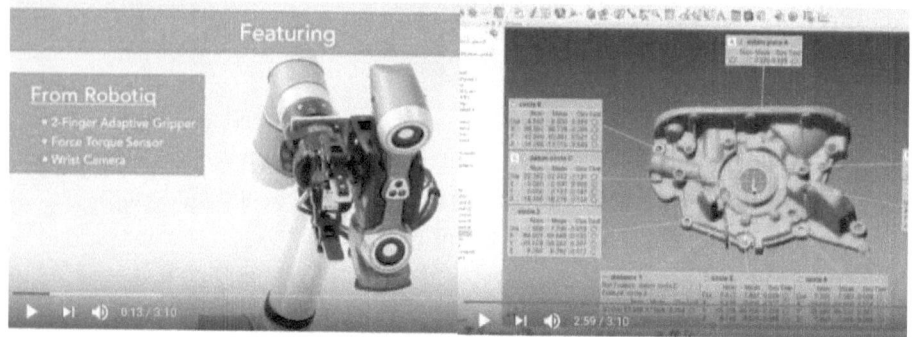

https://youtu.be/38EImpYb1Dw

(Source: YouTube, Robotiq)

To show that the software can be learned by laymen, a video of SensoPart (the employee holds the teach panel of Universal Robots in his hand):

Die Messspitze wird an den Greifpunkt des Bauteils gefahren. Die Position kann gespeichert werden, nachdem der Roboter an den Wegpunkt gebracht wurde.

SENSOPART

https://youtu.be/wRkVEGGq6iY

(Source: YouTube, SensoPart Industriesensorik GmbH)

It is important to know that "optics" is a broad term. In addition to the optics known from photography, which represent what is seen in reality, the use of thermography (thermal image), ultrasound, laser and others can be useful. BMW, for example, has recently started using X-rays to analyze its prototypes. Although not related to robots, this example shows the wide spectrum of possibilities. In the end, almost anything is possible. For many industries and applications there are special solutions which are not presented here individually. For the material carbon (CFRP), for example, Vis-Check offers an optical measurement of the processing quality that can take the individual parameters of the customer into account.

However, for the classic medium-sized company that is getting into the subject, a plug & play solution should be sufficient. He should reserve the individual solution for himself for the next step or however companies that can scale powerfully with a development.

mobility aids

For newcomers to robotics, this is actually not an issue, as it involves additional costs that are actually too high for "getting a taste" of the topic. However, anyone who can imagine moving robots - whether on carriages or rails - should consider in advance whether the robot should also work during movement. Not everyone can do this. For an SME, it can be useful to mount the robot on a mobile stand in order to easily move it to the side if, for example, it is only to be used at night. Such stable carriages are available for about 3.000 €:

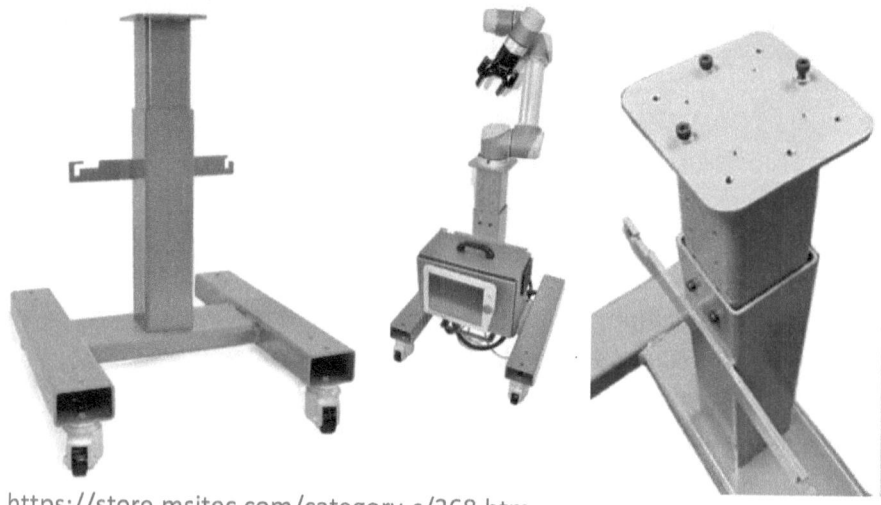

https://store.msitec.com/category-s/368.htm

(Source: MIS Tec)

A larger selection is available from FAUDE in Germany under the name "Provisioning Systems".

If the reach of the Cobot is to be increased or an MRK is to serve two workstations at the same time, a horizontal rail can be an inexpensive option - provided that the robot supports this. This video shows the simple but effective approach:

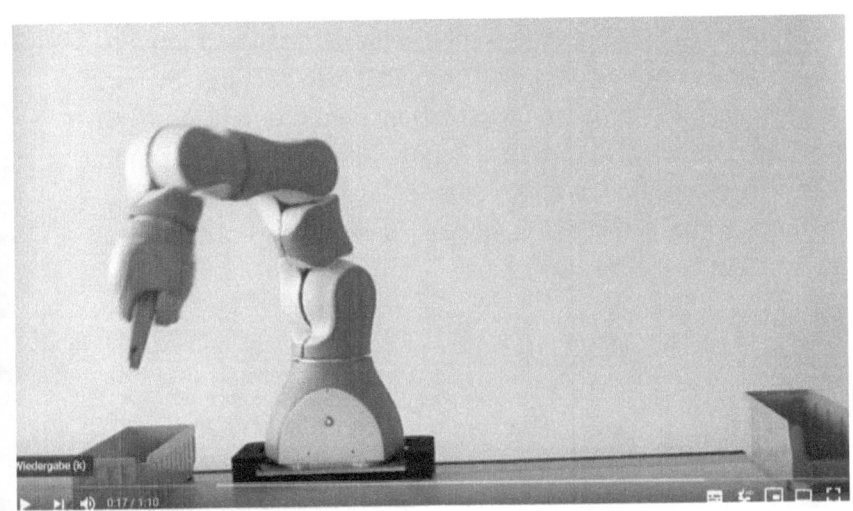

https://youtu.be/FTLza95McMY

(Source: YouTube, F&P Personal Robotics)

In addition to horizontal movement (e.g. using a linear unit), vertical movement is also possible, which is particularly advantageous when palletizing. The Cobot can stack loosely over 2 m height:

https://youtu.be/VytTqE_ODmg

(Source: YouTube, SKF Motion Technologies)

It can also be advantageous from the point of view of the task to buy autonomously moving cars and Cobot from the same manufacturer. By the end of the year there will be autonomous driving solutions for Omron, Kuka, Yaskawa and probably also Universal Robots. The Omron vehicle can carry 60 kg and is also compatible with other Cobots (here in the video a UR). The higher payload makes sense, because beside the MRK its control and maybe material has to be transported.

https://youtu.be/yHWG8Pcs7aY

(Source: You Tube, Axix New York).

The mobile Omron robot travels about 1 hour through its area before its first use to get to know it. Obstacles are detected. Together with a Cobot, a hybrid Cobot is created. The "marriage" of both systems (mobil + Cobot) will be presented in autumn 2019. The travel time with one charge can be about 10 hours, charging time about 3 hours. The mobility surcharge of around € 35,000 is not low, but it can be used to massively increase productivity. A Cobot can, for example, operate three machines instead of one, or move parts in the unmanned shift to where and deposit them there.

Omron makes navigation easier with its landmarks. With the purchase of the Cobot the customer receives landmarks (photo), which he fastens to relevant places.

Each image is photographed once with the camera integrated into the Cobot arm. So the robot learns the next important position. This position can then be assigned a task in the teach panel.

All mobile coasters have one disadvantage: they require a clean, level environment. Caterpillar vehicles or mobile robots with larger rubber tyres do not yet exist in my opinion. This is actually a pity, since companies like Fendt already have the basic knowledge with the "Xaver". The "Xaver" is a small seed robot. His concept is so convincing that he is presented here:

https://youtu.be/a5_kQScrZew

(Source: YouTube, Bayerischer Rundfunk)

The author considers the joint use of such mobile robots with cobots in greenhouses, halls with larger contamination (chips), construction sites or e.g. as harvesting robots conceivable. Harvest Cobots already exist, but they must also be moved.

programming aids

More and more robots are very easy to program. Be it by demonstration or by apps that contain individual program parts. A supplied tablet is often used for this purpose. For the cheap cobots, you must use your own tablet. The risk of compatibility problems is greater here.

If this already simple programming is not enough for you or if you want to use robots that are difficult to program for whatever reason, you can fall back on a tool from the Stuttgart start-up "drag & bot". This Fraunhofer spin-off enables programming even of industrial robots at beginner level. The software costs about 10.000 €.

Cloud

The cloud is also beginning to gain importance in robotics. While the MRK manufacturer Franka Emika has been offering the exchange of programs comparable to iTunes via "FRANKA World" since April 2019, three of the five large software companies Amazon, Microsoft and Google have announced a robot cloud for 2019. This can be used for programming and simulation. The most advanced and well-known are the Amazon considerations. In the Amazon cloud AWS, the Robot Operating System (ROS) can be connected to AWS services. AWS services include machine learning, monitoring and analysis services, and data streaming. It can be navigated and communicated in this way. The entire system is designed to accelerate the processes and protect the processes on site.

As with all cloud considerations, the same applies here: Security relevant or secret does not belong in the cloud! Moreover, consequential costs should not be disregarded. If complex computing operations are relocated to the cloud, monthly costs in the four-digit or even five-digit range can arise if intensive use is made of them.

Protective sheathings

In some industries, such as meat processing, hygiene is a basic requirement. For these cases there are sheaths for robots that can be washed. Conversely, it may be necessary to protect the robot (heavy dust exposure, paint shop). FAUDE, for example, offers suitable "clothing" under the keyword "ProSuit".

Sheaths can also be used for work safety (see below). AirSkin from BlueDanube. This cover is not intended to protect the robot, but to protect the human being. For this purpose, numerous sensors have been incorporated which stop the robot when it hits an object (human). The system is also suitable for making industrial robots more or less collaborative. This means that working side by side is basically possible with AirSkin, but not with each other. Since not every MRK has to have enough sensors, AirSkin can also be of interest for Cobots. This also applies to new models such as those from Universal Robots, as these can move up to 800 mm/sec in collaborative mode with AirSkin. In the video the AirSkin is mounted on a Stäubli, not a real MRK. This reduces the reaction time to 9 ms - the robot's sensors require 40% more time.

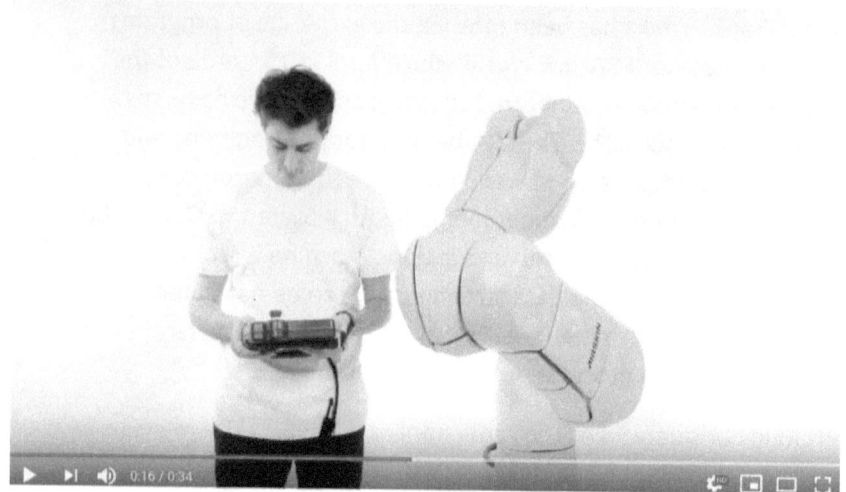

https://youtu.be/AaVpaE-5g-c

(Source: YouTube, Blue Danube Robotics GmbH)

Occupational safety

Even if the MRIs are sensitive, they can still hurt people. A sensitive arm that brakes immediately is of little use if the moving tool can injure the human eye during initial contact. Braking immediately only works at a slower speed or at a speed adapted to the weight. Occupational safety thus has a decisive influence on the productivity of the robot. If he works alone, he can travel at maximum speed and be separated from his human colleagues by light barriers, for example. Companies such as Sick, Keba, Pilz or Mayser have developed appropriate solutions for the coexistence of MRK workers. It is important that not only the robot is evaluated, but also its entire environment. If it has previously been written that trainees can take care of the robot, this does not apply to the important aspect of safety, including the mandatory CE conformity, which must be represented. (The Franka Panda is CE conform "out-of-the-box". Depending on the accessories, however, a new declaration of conformity may be mandatory.) The safety must comply with ISO TS 15066. Universal Robots offers a whitepaper on this subject:

https://www.universal-robots.com/de/suchen/?query=whitepaper#

Franka's PANDA has a valid CE certificate so that it can be used immediately without any further risk assessment when using certain apps and its own grippers. However, in the event of an accident, the question of a situational safety assessment will nevertheless arise. In addition, a documentation must be available formally in one way or another for each workstation.

On the photo of the following video there is a yellow-marked area. If a worker enters, the robot reduces its speed. If the red surface is also entered, it stops altogether. If the employee leaves the areas, geht´s automatically continues. The ideal system thus integrates safety technology and robot control in one control cabinet. If the robot is rarely visited, this effort will not be necessary, then a simple light barrier, which simply stops, can be sufficient.

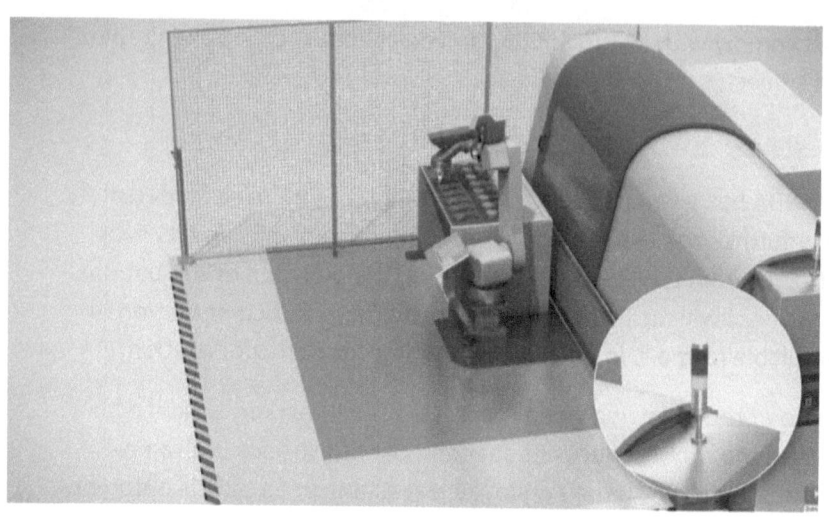

https://youtu.be/yLq9XRy_7ro

(Source: YouTube, Sick Sensor Intelligence)

Pilz is pursuing an alternative approach with its PSENmat safety mat. If an employee enters the mat, the robot is informed (location resolution) and, if necessary, controlled. In addition, the mat has a control function so that the worker can use his leg.

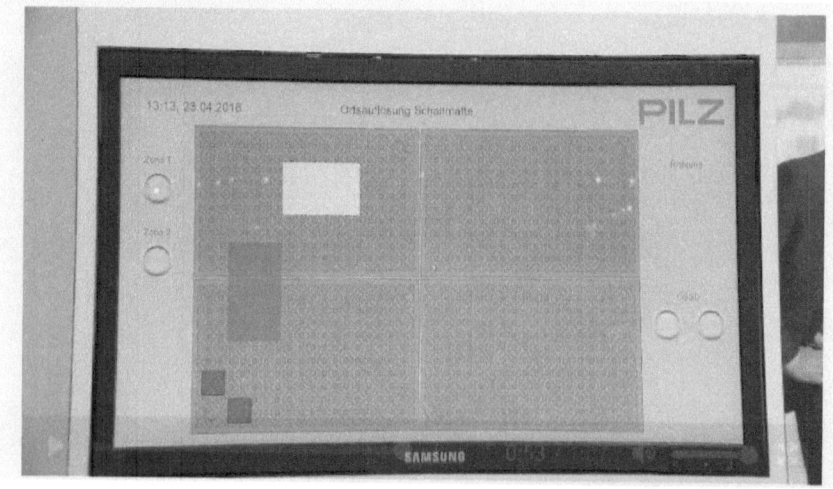

(Source: Mushroom)

Perhaps robots will be equipped with lidar systems in the future, as is the case with autonomous cars. Chinese lidar systems apparently already exist for a smaller four-digit amount. If they work well, the Cobot could detect both obstacles and people at an early stage.

The condition of the robot must be identifiable in terms of occupational safety. This is done by means of lamps, as can be seen. For robots that do not have such a lamp installed, it must be retrofitted. This is not a problem, but it costs money. With Universal Robots, for example, the lamp can be mounted between the arm and the gripper.

The framework conditions are already stricter for cooperation at occasional cooperation / transfer points. Here the question already arises whether the robot, however slowly it moves, can squeeze the operator during operation. The most critical is collaboration. Both work together in the same place - here the strictest safety regulations apply since the risk of unintentional contact is greatest. Additional safety is provided by the Schunk-JL 1 gripper, for example, which has built-in collision detection and can stop the robot.

On the occasion of the Hannover Messe 2019, two innovations were shown which "go beyond" the existing restrictions due to occupational safety. Schunk showed a gripper with a holding force higher than the one previously permitted. Due to a new sensitivity, it slows down the contact with humans so that they cannot be squeezed. The sensor manufacturer Sick again presented a sensor based on 5G. This new, ultra-fast data transmission means that a robot can be stopped later than before, since the buffer time previously required is no longer required.

For a first impression, we recommend the free "KUKA HRC" app. Various parameters such as directional speed, contact area, mass to be moved and endangered body part can be entered in these. The head is consequently protected more strongly than, for example, the thigh. This means that robot activities at head height have a lower maximum speed during collaboration than those at ground level. The app calculates the acceptable speed for each parameter. The pain value table of ISO TS 15066 states a force of 65 N for the face and 220 N for the thigh.

It is possible that the existing occupational safety regulations will be relaxed for Cobots in 2020 to take account of their sensitivity and slower processes.

Complete solutions

More and more companies want not only a Cobot, but a complete solution, so that a high degree of automation is achieved. And what is useful for one company can also be profitable for other companies. Solutions in the sense of scaling are offered more and more as universal complete solutions. Welding or the equipping of machines are to be mentioned as applications with a particularly large number of solutions.

This video shows how easy welding with Cobot is, for example:

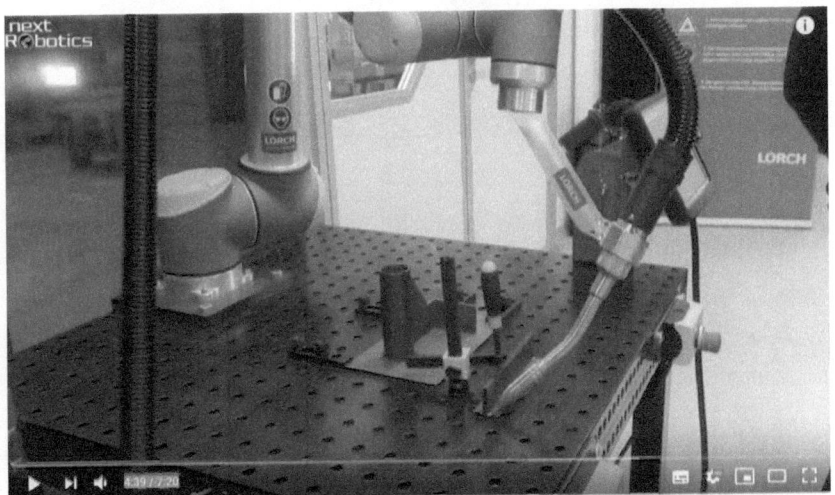

https://youtu.be/vXW3GoanqSw

(Source: YouTube, Next Robotics)

In addition to the Lorch solution shown in the video, there are others, e.g. from Heidenbluth or Invetech. The German-German solution from Panda and VisionLasertechnik is presented here for machine assembly:

https://youtu.be/f5ll5x3lY1c

(Source: YouTube, VisionLasertechnik)

However, almost all other common solutions for machining centres are based on a Universal Robot, whose Payload of up to 10 kg is significantly higher than that of the Panda (3 kg).

Further complete solutions are available for palletizing, grinding/polishing or pick-and-place activities.

Solution to the shortage of skilled workers

At the beginning of July, three companies presented a solution that seemed revolutionary to some and bizarre to others. First of all: As managing director of the participating company VisCheck, the author is intensively involved in the project and therefore certainly not neutral.

What's the point? The Munich start-up VisCheck (image processing using machine learning - AI), the medium-sized company Hufschmied Zerspanungssysteme (technology leader in its field and the first SME to receive an award from BMW for its process engineering) and the Japanese automation and robotics group Omron presented "Opdra". In concrete terms, a camera attached to the robot arm can read out production screens for the first time. This seems bizarre for industry 4.0 followers, as there should be interfaces for this. On older machines these interfaces do not yet exist and on newer machines they are often difficult to create. Even if this is simply possible, "Opdra" still makes sense, since the information - and this is revolutionary - is processed in such a way that corrective action can be taken in the manufacturing process. For this purpose, the robot can program the machine via the keyboard. This in turn appears bizarre as a result of the interfaces mentioned. But also the programming is not yet everything. The intelligence behind it is more important, as are the other options such as machine loading and simultaneous operation of several machines thanks to mobile robots. The animated video shows the final goal of the project. Today, the screen can already be read and easily corrected. The mobile base will be available from autumn 2019; the artificial intelligence, which is already being worked on, and the possibility of ongoing measurement in the production process will be available by mid-2020.

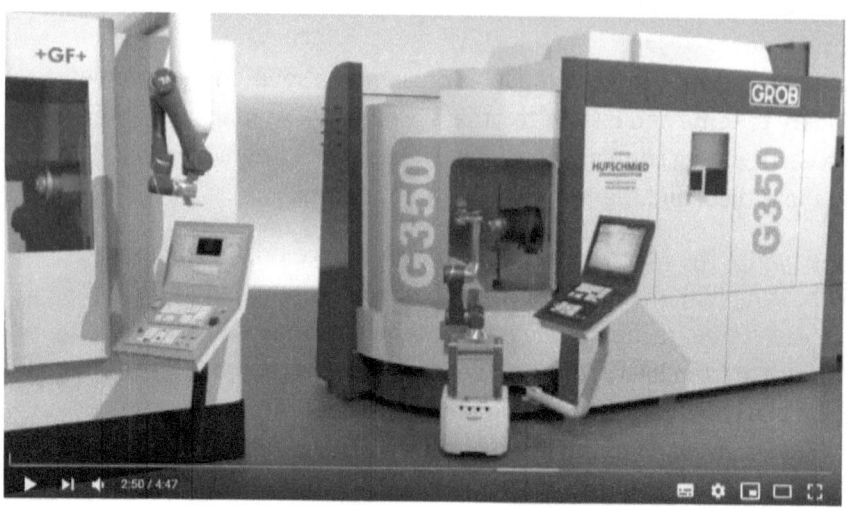

https://youtu.be/jVnlz_qbhDw

(Source: YouTube, VisCheck)

The benefits of the project, which has already received a significant 7-digit amount of development funds, can be described as follows:

1. Unmanned night shifts become less dangerous (committee topic).
2. During the day, the specialist is unburdened.
   3. higher productivity by not interrupting production (sporadic removal of the part for measuring is not necessary).
   4. less / no error parts, as deviations are more likely to be detected and ultimately almost avoided.
   5. higher capacity.
   6 Robot can load/unload parts7
   . Data security - no cloud!

For potential interested parties: The system is already being used by pilot customers and consists exclusively of modules that are compatible with each other and can be retrofitted without any problems. I.e. an early purchase at a lower price has no disadvantages.

After the machining industry, further applications / industries are being processed. The parties involved are open to cooperation. The author also sees application possibilities in the food industry, for example.

The concrete use looks like this, whereby the video was recorded during a public demonstration, which is why the speed of the Cobot was significantly reduced:

https://youtu.be/CJoGssW2XXw

(Source: YouTube, VisCheck)

While some consider the transfer of human knowledge to a Cobot to be impossible, others, e.g. the robot pope Dieter Faude, refer to the system as "Industry 5.0".

Amortization calculation

Instead of a classic amortization calculation, which is already part of the craft of entrepreneurs, a few considerations:

1. For a rough initial cost estimate, it is best to increase the price of the desired robot by the factor 1.5. This should cover the gripper, work safety time and any programming costs.

2. The typical service life for premium suppliers such as Universal Robots is 30,000 hours, while Franka assumes 20,000 hours for its panda. One year contains 8,760 hours at full operation, but full operation is rather rare among the target group of the book. You should therefore reckon with fewer hours. However, this only applies if, for example, the robot is only to be used at night from Monday to Friday and then reaches just under 2,100 hours per year. Because it's unlikely to be used for 15 years. Another aspect, however, would be a permanent extreme load (e.g. maximum load with the arm outstretched).

   If the robot is only used very sporadically, it should nevertheless be taken out of service at some point (after 10 years?). Depending on the requirements, it can therefore correspond to reality to expect only 5,000 hours of service life or only 500 hours per year. But even with such an extremely conservative approach, the hourly rate will still be well below 15 euros, more like 10 euros.

3. The maintenance costs for Cobots are very low and can be set at 5% of the AHK.

4. If the capacity is increased (unmanned night shift), the question arises as to whether the previous sales prices are used or not. If current sales prices are expected, the question arises from a pricing perspective as to why they have not been raised despite capacity bottlenecks. Maybe you can raise the prices.

5. If the amortization calculation turns out to be very good, the question arises whether part of the additional income will not be used for the benefit of a robot or its accessories, which would not actually be needed. In a quiet minute - after a few weeks/months, when

the handling of the MRK is familiar - the next, more complex automation step could be tried out with this equipment. Because robotics should start with something simpler. If the next step is successful, the robot can be used there. For the first, originally conceived solution, it is still possible to buy the cheaper variant. For this consideration, the purchase of a mobile robot holder as well as the selection of a robot suitable for both applications can be useful. The more flexible a Cobot is to be used, the more important is its simple and flexible programming.

6. If there is a promotion, it is nice, but to buy a robot just for the promotion would be wrong. This means that the payback period should also be good without funding.

Epilog - Theses for further development

In the field of robotics, gripper manufacturers seem to be gaining dominance. The accessories will probably become more important than the robot. However, this only applies if the accessories are offered for as many different robot models as possible. The Danish Cluster Odense (headquarters of 120 robotics companies including Universal Robots and OnRobot) with more than 3,200 employees already today, is a danger of concentration to the detriment of the German robotics industry. According to the author, Bavaria and Baden-Württemberg should take countermeasures here.

Globally, robotics is only part of several changes based on technology leaps. With the support of the Bavarian State Government, the Technical University of Munich has concentrated around 30 existing and new chairs for Robotics, Artificial Intelligence and Autonomous Driving in a new institute in the summer of 2018. These three technologies, together with 3D printing, will massively change today's economy. Some theses on this:

1. The new technologies will make it possible to automate numerous workstations within a relatively short period of time. Whether ¼ of all jobs can actually be replaced in Germany is not assessed here.
2. According to the philosopher David Precht, in socio-political terms this should lead to new migratory flows. Because the mixture of 3D printing, autonomously driving trucks and robotics will result in very, very many jobs being lost, especially in Eastern Europe.
3. In Germany, the net loss of jobs is much lower than abroad, as production can be brought home to Germany. Adiddas sets the pace with 3D printing of running shoes. In addition, an industry of manufacturers of the required devices/machines is emerging in Germany.
4. Supported by the negative demographic development, the labour market situation in Germany may remain favourable.
5. New jobs are also likely to be created worldwide, as demonstrated by the transition from agriculture to industrialisation. But this was only true after a few years. In a transitional phase, mass unemployment is definitely threatening. During this time, revolts can

take place when those who are already dependent join forces with those who have been released, according to American professor Richard Baldwin in an interview with the Neue Zürcher Zeitung at the end of 2018.

6. If, however, production facilities move back to Germany and, thanks to automation, more is produced with fewer personnel, the amount of space that is wasted will continue to increase. I.e. less and less people will work on e.g. 100 sqm.

7. As - for example Bavaria - resistance is growing here, in the future industry will also have to build multi-storey buildings, or the obligation to have underground car parks instead of open parking spaces, as is increasingly being discussed in Bavaria. Keller solutions" are also conceivable. In Switzerland, for example, discussions are underway to network all logistics centres underground. CHF 100 million has already been made available for the project planning of the CHF 30 billion project. The change in the framework conditions will consume part of the rationalisation gains.

8. SMEs must be careful, otherwise they will be among the losers of increased flexibility. Because by means of Industry 4.0 and Cobots, larger and larger companies will be able to offer competitive products closer and closer to lot size 1. The existing advantage of flexibility will be lost by SMEs. These could score points with a strong focus on low-cost partial automation. Large companies will have to choose the more expensive full automation.

Contact us

Guido Break

Guido.Bruch@mrk-blog.de

https://www.Mrk-blog.de

EA Enterprise Development GmbH
Merzstr. 16

https://www.equity-advice.de
 D-81679 Munich, Germany

Phone: ++49/ 89 189 378 77-0

https://mrk-blog.de/angebot/